AF302819

Model Order Reduction for Efficient EMC Simulation

Vom Fachbereich 18
Elektrotechnik und Informationstechnik
der Technischen Universität Darmstadt

zur Erlangung des akademischen Grades
eines Doktor-Ingenieurs (Dr.-Ing.)
genehmigte

DISSERTATION

von

Dipl.-Ing. Ghislain Mouil Sil
geboren am 24. Mai 1982 in Douala

Referent: Prof. Dr.-Ing. Thomas Weiland
Korreferent: Prof. Dr. techn. R. Dyczij-Edlinger

Tag der Einreichung: 13.07.2010
Tag der mündlichen Prüfung: 03.11.2010

D17
Darmstadt 2010

Bibliografische Information der Deutschen Nationalbibliothek
Die Deutsche Nationalbibliothek verzeichnet diese Publikation in der Deutschen
Nationalbibliografie; detaillierte bibliografische Daten sind im Internet über
http://dnb.d-nb.de abrufbar.
1. Aufl. - Göttingen: Cuvillier, 2010
Zugl.: (TU) Darmstadt, Univ., Diss., 2010

978-3-86955-559-1

© CUVILLIER VERLAG, Göttingen 2010
Nonnenstieg 8, 37075 Göttingen
Telefon: 0551-54724-0
Telefax: 0551-54724-21
www.cuvillier.de

1. Auflage, 2010
Gedruckt auf säurefreiem Papier

978-3-86955-559-1

Zusammenfassung

In den letzten Jahren ist die Anzahl der elektronischen Komponenten im Auto stetig gestiegen. Sowohl für Sicherheitsaspekte als auch für Multimedia-Lösungen werden immer komplexere Systeme eingesetzt. Die Folge dieser Entwicklung sind große Datenmengen, die mit immer höher werdenden Geschwindigkeiten ausgetauscht werden müssen. Auf der anderen Seite stellt der Trend zur Elektromobilität auch hohe Herausforderungen für die EMV dar. Zur Sicherstellung deren Funktionalität werden restriktive EMV-Normen an die elektronischen Komponenten gesetzt. Die EMV-Simulation hat aufgrund der Fortschritte sowohl auf der Hardware- als auch auf der Software-Seite in den Produktentstehungsprozess Einzug gehalten. Die hohe Komplexität der Komponenten und die Nachfrage nach immer präziseren Vorhersagen stellen aber hohe Anforderungen.

Im Rahmen dieser Arbeit wurde die Modellordnungsreduktion (MOR) zur Effizienzsteigerung der EMV Simulation eingesetzt. Um den Anforderungen insbesondere im Hinblick auf eine Kopplung Feld- und Schaltungssimulation gerecht zu werden, wurde die sogenannte passive Padé Approximation implementiert. Diese Methode ermöglicht schnelle Berechnungen, ist geeignet für resonante Strukturen und liefert passive, genaue und kompakte Modelle. MOR wurde in ein Konzept, dessen Kern ein C++ Code darstellt und, das von der EMV-Vorhersage über die Analyse bis hin zur Optimierung das gesamte Spektrum der EMV Simulation abdeckt, integriert.

Die passive Padé Approximation wurde um eine online-Fehlerkontrolle ergänzt. Dies ist zur Aufstellung eines effizienten Abbruchkriteriums unabdingbar. Durch die Generierung passiver und genauer Makromodelle können Feld- und Schaltungssimulation besser miteinander gekoppelt werden. MOR ermöglicht zusätzlich einen Beschleunigungsfaktor von bis zu 30 verglichen zu herkömmlichen Frequenzbereichsmethoden. Durch die Parallelisierung kann der Code komplexere Strukturen rechnen und somit den Modellierungsaufwand reduzieren und dabei die Genauigkeit der Simulation verbessern. Ein reales Beispiel mit 8.10^6 Unbekannten wurde in 1h30min. auf 48 Prozessoren berechnet, während iterative Löser auf einer 64 GB Maschine nach einer Woche nicht konvergieren konnten.

Die Methode zur EMV-Analyse, die auf die Generierung physikalisch bedingter Ersatzschaltbilder - in Kontrast zu oben genannten Makromodellen, die lediglich eine mathematische Konstruktion ohne physikalischen Bezug sind -, konnte erfolgreich an einem ESP (electronic stabilization program) Modell angewandt werden. Sie besteht darin, die Systeme resultierend aus MOR und aus der Knotenanalyse als rationale Polynome darzustellen. Die Koeffizienten dieser Polynome werden anschließend durch Optimierung abgeglichen. So können inkompatible Schaltungen automatisch eliminiert und die Werte der vorkommenden Elemente ermittelt werden. Dadurch lassen sich die Ursachen parasitischer Effekte effizient identifizieren.

Die EMV-Optimierung besteht aus diskreten Maßnahmen, die die Berechnung verschiedener geometrischen Varianten erfordert. Zu diesem Zweck wurde eine Methode, bei der geometrische Variationen (Versatz von Leitungen oder Bauelementen) direkt an den Systemmatrizen durchgeführt werden, in Kombination mit einem genetischen Algorithmus implementiert. So verläuft der Optimierungsprozess autonom und die Zeit für die Vergitterung (in der Größenordnung von ein paar Minuten pro Iterationsschritt für komplexe Systeme) kann gespart werden. Diese Methode konnte erfolgreich zur Optimierung eines DC-DC Wandlers für Anwendung in Hybridmotoren eingesetzt werden.

Contents

Contents

1 Introduction

1.1 Motivation

1.1.1 Introduction to EMC

Electromagnetic compatibility (EMC) is the ability of a device or system to function without error in its intended electromagnetic environment, without influencing this environment inadmissibly [1]. This is of great importance as transfer of electromagnetic energy induces interferences between electronic devices which could in the context of e.g. automobile safety lead to severe damages.

EMC standards are set today by international CISPR (comité international spécial pour les pertubations radioélectriques) recommendations from which European standards are derived by the European standardization institute for civil applications CENELEC (comité européen de normalisation electrotechnique). The institute which is responsible for the national DIN standard in Germany is the DEK (Deutsches Elektrotechnisches Komitee) [1].

There are two main trends observed in the last years which enhanced the interest on EMC in automobiles:

- In order to meet the demand for telecommunication services and improve the safety standards, faster and more sophisticated electronic systems are provided.

- The implementation of electromobility requires the integration of more electronic devices.

Summing up, those trends lead to more devices and thus a higher electromagnetic susceptibility. In order to guarantee the safety while providing more comfort and possibilities for the passengers, the EMC norms have got more and more restrictive. This progress has contributed to put EMC engineering in the focus of electronic product development.

For EMC validation, it can be distinguished between electromagnetic emissions of the device under test (DUT) and its immunity against interferences from its environment which are related to *external EMC*, whereas the interferences of components within a product or system are related to *internal EMC* [2]. In the scope of this work, the focus will be on internal EMC and electromagnetic emissions. For a detailed insight into this topic, refer to [2–5].

1.1.2 EMC Simulation in this Work

EMC simulation has grown in the last years to an indispensable tool for EMC engineering. In fact, it allows a reduction in time and resources in early design stages as well as later in the development process. Tremendous advances in hardware and computational methods have contributed to this progress. This interest has increased the demand on model accuracy which ramped up again computation time and model complexity. Improving the underlying computational methods is thus of great importance for a better efficiency of EMC simulations.

The electronic devices analyzed in this work are subdivided into two categories:

- **Control units** are impregnated in printed circuit boards (PCBs) which are multi-layered and miniaturized ($\sim \mu$m details). These devices become more and more indispensable especially for transmission control and safety issues in cars[1]. In this work we analyzed the board of an electronic stabilization program (ESP) system.

- **DC/DC-Converters** have gained interest in the last years due to its application in hybrid and electrical cars which constitute the megatrend electromobility. The dimensions of the devices here are larger compared to PCBs ($\sim$mm).

The numerical modeling of these problem types which contain nonlinear elements[2] requires the coupling of three dimensional (3D) field- and circuit simulation. The high frequency (HF) couplings occurring on the PCB or converter are captured by a transfer function computed with a field solver. From the transfer function a reduced model is derived for the circuit simulation, which additionally considers nonlinear elements.

For the discretization in field solvers, we distinguish between

---

[1]They are used for steering, anti-lock bracking, suspension and engine management systems.

[2]Nonlinear in system view, e.g. micro controllers, transistors, ...

- **Surface discretization methods**: In this set of methods (e.g. *Boundary element method, BEM* [6]), only the conducting structures surfaces inside of homogeneous bodies and dielectrica are discretized as the wave propagation in air is compted with help of Green functions.

- **Volume discretization methods**: These methods limit the number of state variables by considering only a part of the whole domain as computational domain in which the field values are solved for nodes defined on a grid. In this regard, the *finite element method (FEM)* approximates the requested functions of the fields through superpositions e.g. of functions which are each defined on small areas. The wave equation can be transformed in the following system [7]

$$(\mathbf{A} + \frac{\mathrm{d}}{\mathrm{d}t}\mathbf{D} + \frac{\mathrm{d}^2}{\mathrm{d}t^2}\mathbf{K})\mathbf{e} = \mathbf{b}, \qquad (1.1.1)$$

  where $\mathbf{e}$ is the unknown field vector and $\mathbf{b}$ the excitation vector. The stiffness matrix $\mathbf{A}$, the damping matrix $\mathbf{D}$, and the mass matrix $\mathbf{K}$ are all sparse. On the other side, the *finite differences (FD)* computes the fields on each node where the partial differential form of Maxwell's equations is transformed in discrete differences. The finite integration technique (FIT) which on its part consists on applying the integral form of the Maxwell's equations on the the nodes of the grid can also be seen as part of this group.

In the scope of this work, volume based techniques are more suitable, especially considering PCBs as volume elements (air, dielectrics, ...) dominate surface elements (traces, layers, ...). The volume discretization of PCBs yields generally 10^6 to 10^7 unknowns whereas only 10^5 unknowns are needed for converters. We used FIT as discretization method in this work, but it should be stated that the methods implemented can be easily extended to FEM systems.

After having discretized the DUT, the computation of its transfer function resulting from the Maxwell equations may be performed with one of the following methods:

- Time domain: This method has a low complexity as only matrix-vector multiplications[3] are performed. It is appropriate for problems with wide frequency range but gets less inefficient for resonant structures[4] and in presence of a lot of ports[5] [8].

---

[3] Only in combination with FIT or FD which allow efficcient explicit time domain methods.
[4] The energy in the system would decay slower and thus lead to longer computation times.
[5] The computation time grows linearly with the number of ports, except in combination with Graphic process units (GPU)

- Frequency domain: The complexity of this method is high as several matrix-inversions should be performed to determine the transfer function in a wide frequency range. However, it may get more efficient as time domain in presence of a lot of ports[6] and for narrow band computations.

The main challenges for field computation for EMC purposes can be stated as follows:

- wide frequency range,

- large model size,

- resonant behavior,

- and large number of ports.

The limits in circuit simulation are:

- accuracy of extracted models,

- number of ports,

- guarantee for passivity and stability.

1.2 MOR

Model order reduction methods (MOR) [9] which consist on computing reduced order models have been first used in the field of control theory. They have been since then introduced to systems resulting from nodal analysis and discretization of Maxwell's equations. They generally consist on generating a reduced model which captures the dependance of a function on one or several parameters. By this way, the complexity of recomputations is lowered and the coupling with other models is made easier.

MOR methods which have already been identified as robust for efficient field simulation [10–12] and especially for EMC simulation [13] present the following advantages:

- fast computation,

- wide frequency range,

- efficient for resonant structures,

---

[6]Once the matrix has been inverted, it can be applied at all ports.

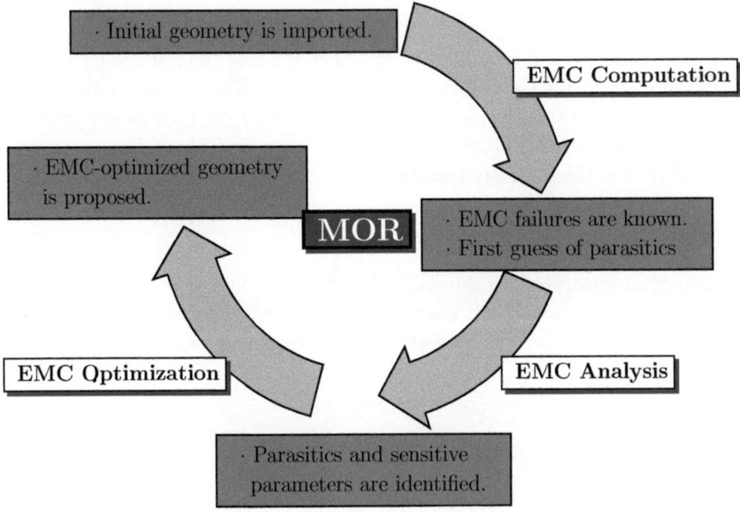

Figure 1.1: Overview of the EMC simulation concept implemented in the scope of this work

- high accuracy of the macromodels,

- guaranteed passivity of extracted models.

MOR has been implemented in a whole concept from EMC computation (prediction) through EMC analysis to EMC optimization as illustrated in Figure 1.1. This concept guarantees a better integration of EMC simulation in the development process of electronic devices.

EMC Computation

In this step, the EMC behavior of a given structure is predicted through simulation. It involves the coupling of field and circuit simulation already discussed above. MOR can be used to compute the transfer function more efficiently and/or to generate passivity preserving models for a stable circuit simulation if some system currents or voltages are of interest.

EMC Analysis

If some EMC failures have been predicted in the first step, an analysis is indispensable in order to understand their causes and thus derive proper EMC measures for improvement. In this step, we make use of the exact mathematical expression of the transfer function with MOR in order to generate equivalent circuits which retrieve the physics of the considered structure.

EMC Optimization

The EMC measures which typically influence the structure geometry are then parameterized and combined with a suitable optimization algorithm to retrieve an EMC-optimized geometry. Again, the efficiency of MOR can be used to reduce the overall optimization time.

1.3 Outline

After this introduction, the method FIT will be presented in Chapter 2. First, the discretization of Maxwell's equations will be addressed and afterwards, the discretized models will be derived to state space representations whose system properties are discussed.

The topic of Chapter 3 is model order reduction. After a brief introduction, different MOR methods are presented. Especially its passivity preserving variant which has been implemented in this work will be addressed.

The main contribution of this work is presented in Chapter 4 where the concept introduced in Section 1.2 is explicitly explained. The main concern in the EMC computation part is the efficient implementation of MOR. Particularly, an error control method for a reliable stop criterion is presented. A method for efficient order reduction in presence of a high number of ports is also proposed. Furthermore, an appropriate parallelization technique to enable the computation of more complex structures is introduced. The next section on EMC analysis presents a method to generate equivalent physical circuits by matching of the polynomial representations of the transfer functions resulting from MOR to proposed circuits. In the section related to EMC optimization, the computation of different variants is described along with the optimization workflow implemented in this work.

In Chapter 5, a method of matrix compression, the so-called Kronecker decomposition, is presented. Differently from parallelization which enables the computation of complex structures on several machines, this method allows to compute models of high order (10^7 unknwons) on standard computers with less memory requirements. It should be stated that the method requires cartesian grids and is thus suitable for FIT. On the other hand, this method has not been implemented in this work, as there are still some open issues to address on this field.

The methods implemented in this work are applied to solve real-life problems which are presented in Chapter 6. They consist of a measurement setup of the electronic stabilization program (ESP), and a DC/DC-converter.

Finally, Chapter 7 gives a conclusion to this work and an outlook to further improvements.

2 Finite Integration Technique

Maxwell's equations are the constitutive equations for the description of electromagnetic fields. Therefore, they should be considered for the EMC analysis of electronic devices. The numerical computation requires the discretization of those equations in the considered domain. It can be distinguished between surface and volume discretization methods. For our purposes, latter have been identified as the most suitable. Among them, the finite element method, finite differences and the finite integration are the most popular numerical methods for the solution of Maxwell's equations.

In this work, we implemented the finite integration technique on orthogonal grids (hexaeder elements) because of the easiest mesh generation and the possibility to apply the matrix compression introduced in Chapter 5. Finite element methods have surely a better convergence behavior since they can make use of basis functions of higher order, but this comes at cost of a more complex computation. Furthermore while the ratio of unknowns between orthogonal and unstructured grids for PCB simulations (details in the order of 100 μm) is very low, it tends to one for structures in power electronics.

In this chapter we recall first of all the Maxwell's theory and its discretization with FIT. As the transfer function is of great interest for EMC analysis, the FIT systems are presented and analyzed. While the eigenvalue analysis is essential for a proper handling of the singular differential operators of the wave equation, the system properties are indispensable for a coupling with transient simulations.

2.1 Maxwell's Equations

Ampere (1826) and Faraday (1831) put the coupling between electric and magnetic fields with the Ampere's and the induction laws in evidence. However, the four constitutive equations of classical electrodynamics are related to Maxwell because he revealed the bidirectional coupling of electric and magnetic fields and was the precursor of the today well known uniform electromagnetic theory [14].

Maxwell's equations connect the electric field $\vec{E}$, the electric flux density $\vec{D}$, the current density $\vec{J}$, the magnetic field $\vec{H}$, the magnetic flux density $\vec{B}$ and the charge density ρ. The first two equations - also called Faraday's and Ampere's laws - show that the time derivative of a flux over a given surface A is equal to the voltage of its boundary ∂A. Whereas the next two equations show the equivalence between the charge in a volume V and the flux over its closed boundary ∂V express the nonexistence of magnetic charge. For more details, we recommend [15]. Assuming static media, Maxwell's equations are given the following way

$$\oint_{\partial A} \vec{E}(\vec{r},t) \cdot \mathrm{d}\vec{s} = -\int_A \frac{\partial \vec{B}(\vec{r},t)}{\partial t} \cdot \mathrm{d}\vec{A} \tag{2.1.1a}$$

$$\oint_{\partial A} \vec{H}(\vec{r},t) \cdot \mathrm{d}\vec{s} = \int_A \left(\frac{\partial \vec{D}(\vec{r},t)}{\partial t} + \vec{J}(\vec{r},t) \right) \cdot \mathrm{d}\vec{A} \tag{2.1.1b}$$

$$\oint_{\partial V} \vec{D}(\vec{r},t) \cdot \mathrm{d}\vec{A} = \int_V \rho(\vec{r},t)\mathrm{d}V \tag{2.1.1c}$$

$$\oint_{\partial V} \vec{B}(\vec{r},t) \cdot \mathrm{d}\vec{A} = 0. \tag{2.1.1d}$$

The current density in (2.1.1b) is given as the sum

$$\vec{J}(\vec{r},t) = \vec{J}_e(\vec{r},t) + \vec{J}_c(\vec{r},t) + \vec{J}_l(\vec{r},t) \tag{2.1.2}$$

of an externally impressed current density, $\vec{J}_e(\vec{r},t)$, the conducting current density, $\vec{J}_l(\vec{r},t)$, induced by the conductivity in materials and the convection current density, $\vec{J}_c(\vec{r},t)$, induced by moving charges due to electromagnetic forces.

The Gauss' and Stokes' laws allow a transformation of Maxwell's equations into their differential form

$$\mathrm{curl}\ \vec{E}(\vec{r},t) = -\frac{\partial \vec{B}(\vec{r},t)}{\partial t}, \tag{2.1.3a}$$

$$\mathrm{curl}\ \vec{H}(\vec{r},t) = \frac{\partial \vec{D}(\vec{r},t)}{\partial t} + \vec{J}(\vec{r},t), \tag{2.1.3b}$$

$$\mathrm{div}\ \vec{D}(\vec{r},t) = \rho(\vec{r},t), \tag{2.1.3c}$$

$$\mathrm{div}\ \vec{B}(\vec{r},t) = 0. \tag{2.1.3d}$$

The vectorial variables need to satisfy the continuity condition at boundaries between electrically or magnetically different media for the differential form to be correct. The conditions are given as

$$\vec{n}_{12} \times (\vec{E}_1 - \vec{E}_2) = 0,\ \vec{n}_{12} \times (\vec{H}_1 - \vec{H}_2) = \vec{J}_F, \tag{2.1.4a}$$

$$\vec{n}_{12} \cdot (\vec{D}_1 - \vec{D}_2) = \sigma_F,\ \vec{n}_{12} \cdot (\vec{B}_1 - \vec{B}_2) = 0, \tag{2.1.4b}$$

with $\vec{n}_{12}$ being the normal vector to the boundary surface, σ_F the surface charge density and $\vec{J}_F$ the surface current density.

2.1.1 Material Properties

Maxwell's equations can only be solved when apart from (2.1.1) or (2.1.3) the material properties which establish the relationship between field strengths and flux densities are known. For the general, inhomogeneous and anisotropic case, they are given as

$$\vec{D}(\vec{r},t) = \varepsilon_0 \vec{E}(\vec{r},t) + \vec{P}(\vec{E},\vec{r},t), \tag{2.1.5a}$$

$$\vec{B}(\vec{r},t) = \mu_0 \vec{H}(\vec{r},t) + \mu_0 \vec{M}(\vec{H},\vec{r},t), \tag{2.1.5b}$$

$$\vec{J}_l(\vec{r},t) = \sigma \vec{E}(\vec{r},t). \tag{2.1.5c}$$

Thus, the flux consists of a linear dependence on the field strength in vacuum and a in general complex influence of the material defined through the polarization $\vec{P}$ and the magnetization $\vec{M}$. In this way, dispersive, anisotropic, nonlinear and frequency dependent effects can be described. In most cases, we can assume linear homogenous and isotropic materials, so that (2.1.5a) and (2.1.5b) are turned into

$$\vec{D}(\vec{r},t) = \varepsilon_0 \varepsilon_r \vec{E}(\vec{r},t), \qquad \varepsilon_r \in \mathbb{R}, \tag{2.1.6a}$$

$$\vec{B}(r',t) = \mu_0 \mu_r \vec{H}(\vec{r},t), \qquad \mu_r \in \mathbb{R}. \tag{2.1.6b}$$

2.2 Discretization of the Maxwell's Equations

Maxwell's Equations and the material properties described above allow the exact prediction of the magnetic and electric fields and fluxes assuming all sources are known. However, analytic solution of these equations is only possible on more or less trivial structures or after some simplifications.

Numerical methods allow to solve the Maxwell's equations approximatively by modeling the continuous fields through a given number of state variables. They thus make it possible to handle real structures. There are two main methods in this respect:

- Volume based techniques: The finite integration technique (FIT) related to finite differences (FD), and finite element method (FEM) are the most used for Maxwell's equations.

- Surface based techniques such as the boundary element method (BEM).

The finite integration technique (FIT) first presented in 1977 by T. Weiland in [16] consists of applying the integral form of Maxwell's equations (2.1.1) directly on the elements of the meshed computational domain. Thus, the physical properties of the continuous fields are kept in the discrete form. The following is oriented on [17–20].

2.2.1 Maxwell's Equations on a Grid

The numerical computation by the FIT is based on an appropriate discretization of the domain of concern which allows a good approximation of the structure[1]. For this purpose, different types of grids may be defined, from general nonorthogonal grids, tetrahedral grids to Cartesian or cylindrical grids. In this work, we consider orthogonal Cartesian grids as illustrated in Fig. 2.1.

The domain is first of all discretized in a Cartesian 3D-grid (G). For efficiency reasons, the cells are numerated following the coordinate directions. With I, J, and K being the number of points in the 3 directions and i, j, and k the indices, the $N_p = I \cdot J \cdot K$ grid points are numbered as follows:

$$n(i, j, k) = i + (j - 1)I + (k - 1)IJ \tag{2.2.1}$$

With the electric voltage $\widehat{e}_p$ ($p \in \{u, v, w\}$) defined as integral of $\vec{E}$ over the edge length L_p and the magnetic flux $\widehat{\widehat{b}}_p$ as surface integral of $\vec{B}$ over the surface A_p,

$$\widehat{e}_p(i, j, k) = \int_{L_p(i,j,k)} \vec{E} \cdot \mathrm{d}\vec{s}, \quad \widehat{\widehat{b}}_p(i, j, k) = \int_{A_p(i,j,k)} \vec{B} \cdot \mathrm{d}\vec{A}, \tag{2.2.2}$$

(2.1.1a) applied on the surface $A_w(i, j, k)$ leads to

$$\widehat{e}_u(i, j, k) + \widehat{e}_v(i + 1, j, k) - \widehat{e}_u(i, j + 1, k) - \widehat{e}_v(i, j, k) = -\frac{\mathrm{d}}{\mathrm{d}t} \widehat{\widehat{b}}_w(i, j, k) \tag{2.2.3}$$

This description is exact because of the use of integral values. If $\widehat{\mathbf{e}}$ is the vector containing all the electric voltages[2] and $\widehat{\widehat{\mathbf{b}}}$ the one containing all the magnetic fluxes then (2.2.3) can be transformed for all the surfaces in the domain to a system of equations:

$$\mathbf{C}\widehat{\mathbf{e}} = -\frac{\mathrm{d}}{\mathrm{d}t}\widehat{\widehat{\mathbf{b}}} \tag{2.2.4}$$

The matrix $\mathbf{C}$ ($3N_p \times 3N_p$) has the same meaning like the *curl − operator* in (2.1.3a). It has only two nonzero elements $(1, -1)$ per row, is highly sparse and singular. As $\widehat{\mathbf{e}}$

---

[1]The fineness of the mesh depends not only on the structure details but also on the frequency of the fields.

[2] $\widehat{\mathbf{e}}$ has $3N_p$ entries, N_p for each direction.

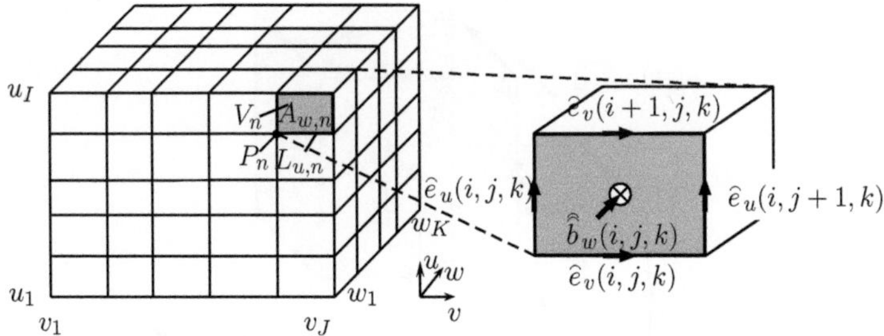

Figure 2.1: Discretization of a structure through a Cartesian grid. Electric voltages and magnetic flux for the surface $A_w(i, j, k)$

and $\overset{\approx}{\mathbf{b}}$ in (2.2.4) are sorted from the u-, then v- and at least the w-direction, $\mathbf{C}$ has a band structure. With

$$[\mathbf{P}_{u,v,w}]_{[p,q]} = \left\{ \begin{array}{lll} -1 & : & p = q \\ 1 & : & p = q + r \\ 0 & : & \text{else} \end{array} \right\} \tag{2.2.5}$$

and $\mathbf{P}_u : r = 1$, $\mathbf{P}_v : r = I$ and $\mathbf{P}_w : r = I \cdot J$, $\mathbf{C}$ can be written as below:

$$\mathbf{C} = \left(\begin{array}{ccc} 0 & -\mathbf{P}_w & \mathbf{P}_v \\ \mathbf{P}_w & 0 & \mathbf{P}_u \\ -\mathbf{P}_v & \mathbf{P}_u & 0 \end{array} \right). \tag{2.2.6}$$

Also, (2.1.1d) can be discretized the same way and yields:

$$\mathbf{S}\overset{\approx}{\mathbf{b}} = 0 \quad \text{with } \mathbf{S} = (\mathbf{P}_u, \quad \mathbf{P}_v, \quad \mathbf{P}_w). \tag{2.2.7}$$

The matrix $\mathbf{S}$ is equivalent to the *source − operator* in (2.1.1d). The remaining Maxwell equations are discretized on the so-called *dual grid* which is orthogonal to the original one (Fig. 2.2). Each edge of the dual grid intersects at right angle a surface of the primary grid (and vice-versa) and each cell of one of the two grids contains a node of the other one.

With the vectors $\overset{\frown}{\mathbf{h}}$, $\overset{\approx}{\mathbf{b}}$, $\overset{\approx}{\mathbf{j}}$, and $\mathbf{q}$ and the corresponding operators $\widetilde{\mathbf{C}}$ and $\widetilde{\mathbf{S}}$ we can define the grid-equations as follows:

$$\mathbf{C}\overset{\frown}{\mathbf{e}} = -\frac{d}{dt}\overset{\approx}{\mathbf{b}}, \tag{2.2.8a}$$

$$\widetilde{\mathbf{C}}\overset{\frown}{\mathbf{h}} = \frac{d}{dt}\overset{\approx}{\mathbf{d}} + \overset{\approx}{\mathbf{j}}, \tag{2.2.8b}$$

$$\widetilde{\mathbf{S}}\overset{\approx}{\mathbf{d}} = \mathbf{q}, \tag{2.2.8c}$$

$$\mathbf{S}\overset{\approx}{\mathbf{b}} = 0. \tag{2.2.8d}$$

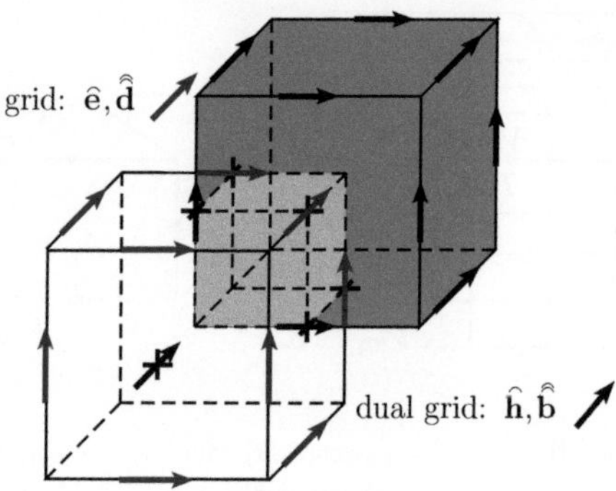

Figure 2.2: Representation of the dual grid relative to the primary grid and the corresponding voltages and fluxes.

The dual relation $\mathbf{C}^T = \widetilde{\mathbf{C}}$ is an essential property of this discretization. As stated before, the governing analytic relations are available for the discretized system

$$\mathbf{SC} = \widetilde{\mathbf{S}}\widetilde{\mathbf{C}} = \mathbf{0} \iff \text{div rot} \equiv 0 \tag{2.2.9a}$$

$$\mathbf{C}\widetilde{\mathbf{S}}^T = \widetilde{\mathbf{C}}\widetilde{\mathbf{S}}^T = \mathbf{0} \iff \text{rot grad} \equiv 0, \tag{2.2.9b}$$

which makes this method consistent.

2.2.2 Material Discretization

The definition of the material equations on the grid introduces some approximations which are unavoidable in numerical methods. They relate the flux and voltage values with each other and thus the primary and dual grid. Following (2.1.5) and assuming linear materials with linear polarization, the grid-material equations are given as

$$\widehat{\widehat{\mathbf{d}}} = \mathbf{M}_\varepsilon\,\hat{\mathbf{e}}, \tag{2.2.10a}$$

$$\widehat{\widehat{\mathbf{j}}} = \mathbf{M}_\sigma\,\hat{\mathbf{e}} + \widehat{\widehat{\mathbf{j}}}_e, \tag{2.2.10b}$$

$$\hat{\mathbf{h}} = \mathbf{M}_\mu^{-1}\widehat{\widehat{\mathbf{b}}}, \tag{2.2.10c}$$

where $\widehat{\widehat{\mathbf{j}}}_e$ represents the impressed currents. In the case of dual orthogonal grids, each voltage value is associated to a flux value so that the matrices $\mathbf{M}_\varepsilon$, $\mathbf{M}_\sigma$ and $\mathbf{M}_\mu^{-1}$ are diagonal. Assuming that field strengths and flux densities are constant over the corresponding lengths and surfaces, the material matrices are given as

$$\mathbf{M}_{\varepsilon[p,p]} = \frac{\int_{\tilde{A}_p} \varepsilon \, dA}{L_p} \rightarrow \mathbf{M}_\varepsilon = \tilde{\mathbf{D}}_A \mathbf{D}_\varepsilon \mathbf{D}_S^{-1}, \tag{2.2.11a}$$

$$\mathbf{M}_{\sigma[p,p]} = \frac{\int_{\tilde{A}_p} \sigma \, dA}{L_p} \rightarrow \mathbf{M}_\sigma = \tilde{\mathbf{D}}_A \mathbf{D}_\sigma \mathbf{D}_S^{-1}, \tag{2.2.11b}$$

$$\mathbf{M}_{\mu[p,p]}^{-1} = \frac{\int_{\tilde{L}_p} \mu^{-1} \, ds}{A_p} \rightarrow \mathbf{M}_\mu^{-1} = \tilde{\mathbf{D}}_S \mathbf{D}_\nu \mathbf{D}_A^{-1} = \mathbf{M}_\nu \tag{2.2.11c}$$

where $\mathbf{D}_A$ and $\mathbf{D}_S$ are metric matrices and $\mathbf{D}_\varepsilon$ and $\mathbf{D}_\nu$ ($\nu = 1/\mu$) contain the permittivity and inverse permeability values, respectively. As the mesh cells are homogenously filled, the discretization of structures with fine details may lead to a higher number of degrees of freedom. However, some improvements have been already made in this respect with the perfect boundary approximation (PBA) [21] and thin sheet technique (TST) [22].

2.2.3 Boundary Conditions and Excitations

Boundary conditions are indispensable to describe the fields at the boundaries in order to model the impact of the region outside of the computational domain. In this respect, we distinguish between ideal boundaries like electric or magnetic boundaries where no energy interaction with the region outside is considered and other like open boundaries and ports.

Electric Boundary Conditions

Electric boundaries assume an infinite electric conductivity so that the tangential electric voltages and the normal magnetic fluxes vanish. The corresponding surfaces become superfluous and the associated lines and columns of $\mathbf{C}$ and $\mathbf{S}$ can be eliminated. Alternatively, in order to keep the band structure of the matrices, the corresponding entries of the matrices $\mathbf{M}_\varepsilon$ and $\mathbf{M}_\mu^{-1}$ are set to zero[3]. Electric boundaries can be used to simulate structures in conductive housings or as symmetric conditions.

Magnetic Boundary Conditions

Magnetic boundaries assume an infinite magnetic conductivity which lead to vanishing tangential magnetic voltages and normal electric fluxes. Again this can be achieved either by not considering the corresponding edges and surfaces for the operators $\tilde{\mathbf{C}}$

---

[3]It should be stated that matrices are no more invertible. Instead, pseudo-inverses should be computed where only values different from zero are inverted.

and $\widetilde{\mathbf{S}}$ or by setting the associated entries of $\mathbf{M}_\varepsilon$ and $\mathbf{M}_\mu^{-1}$ to zero. In contrast to electric boundaries, this condition does not have any physical meaning except for high permeable materials. It also plays an important role as symmetric condition on structures with symmetric field distribution.

Open Boundaries

The advent of the so-called perfectly matched layer (PML) boundary condition in 1994 [23] was a turning point in simulating open boundaries. Until then, Mur boundaries which are indeed easy to implement but are less accurate were used instead [24]. The PML method consists in adding layers of appropriate nonphysical media to the boundary of the computational domain so that waves can be absorbed with marginal reflections. Thus, an ideal PML absorber has two main properties:

- Refectionless matching to the computational domain for any wave angle and frequency,

- attenuation of the waves in the PML medium.

Several PML techniques have been presented so far. The implemented formulation in this work was first introduced in 1996 [25]: Generalized PML Theory (GT-PML).

The basis of this method lays on Maxwell's equations for an anisotropic medium with complex material properties

$$\nabla \times \mathbf{E} = -j\omega\mu\underline{\mathbf{\Lambda}} \cdot \mathbf{H}, \tag{2.2.12a}$$
$$\nabla \times \mathbf{H} = j\omega\varepsilon\underline{\mathbf{\Lambda}} \cdot \mathbf{E}, \tag{2.2.12b}$$
$$\nabla \cdot (\varepsilon\underline{\mathbf{\Lambda}} \cdot \mathbf{E}) = 0, \tag{2.2.12c}$$
$$\nabla \cdot (\mu\underline{\mathbf{\Lambda}} \cdot \mathbf{E}) = 0. \tag{2.2.12d}$$

with the complex diagonal matrix

$$\underline{\mathbf{\Lambda}} = \begin{pmatrix} \lambda_u & 0 & 0 \\ 0 & \lambda_v & 0 \\ 0 & 0 & \lambda_w \end{pmatrix}, \tag{2.2.13}$$

and $\mathbf{E}$ being the field vector. In this work, the system analysis has been performed in frequency domain where we assume all signals to be harmonic. In this way, the differentiation $\frac{\mathrm{d}}{\mathrm{d}t}$ can be replaced by the term $j\omega$. The matrix $\underline{\mathbf{\Lambda}}$ is dimensionless and causes the spilt of the material properties $\underline{\varepsilon}$ and $\underline{\mu}$ in the PML media in order to achieve wave propagation at any incident angle without reflection. The scale matrix **G** is introduced in order to derive the dispersion equation for the considered medium,

$$(\mathbf{G}\nabla) \times \hat{\mathbf{E}} = -j\omega\mu\hat{\mathbf{H}}, \tag{2.2.14a}$$

$$(\mathbf{G}\nabla) \times \hat{\mathbf{H}} = j\omega\varepsilon\hat{\mathbf{E}}, \tag{2.2.14b}$$

$$(\mathbf{G}\nabla) \cdot (\varepsilon\hat{\mathbf{E}}) = 0, \tag{2.2.14c}$$

$$(\mathbf{G}\nabla) \cdot (\mu\hat{\mathbf{H}}) = 0, \tag{2.2.14d}$$

with

$$\mathbf{G} = \begin{pmatrix} \dfrac{1}{\sqrt{\lambda_v\lambda_w}} & 0 & 0 \\ 0 & \dfrac{1}{\sqrt{\lambda_u\lambda_w}} & 0 \\ 0 & 0 & \dfrac{1}{\sqrt{\lambda_u\lambda_v}} \end{pmatrix}, \ \hat{\mathbf{E}} = \mathbf{GE} \text{ and } \hat{\mathbf{H}} = \mathbf{GH}. \tag{2.2.15}$$

In fact, (2.2.12a) can be derived to (2.2.14a) in cartesian coordinates the following way:

$$\begin{pmatrix} \dfrac{1}{\sqrt{\lambda_u\lambda_w}}\dfrac{\partial}{\partial v}\dfrac{E_w}{\sqrt{\lambda_u\lambda_v}} - \dfrac{1}{\sqrt{\lambda_u\lambda_v}}\dfrac{\partial}{\partial w}\dfrac{E_v}{\sqrt{\lambda_u\lambda_w}} \\ \dfrac{1}{\sqrt{\lambda_u\lambda_v}}\dfrac{\partial}{\partial w}\dfrac{E_u}{\sqrt{\lambda_v\lambda_w}} - \dfrac{1}{\sqrt{\lambda_v\lambda_w}}\dfrac{\partial}{\partial u}\dfrac{E_w}{\sqrt{\lambda_u\lambda_v}} \\ \dfrac{1}{\sqrt{\lambda_v\lambda_w}}\dfrac{\partial}{\partial u}\dfrac{E_v}{\sqrt{\lambda_u\lambda_w}} - \dfrac{1}{\sqrt{\lambda_u\lambda_w}}\dfrac{\partial}{\partial v}\dfrac{E_u}{\sqrt{\lambda_v\lambda_w}} \end{pmatrix} = -j\omega\mu\begin{pmatrix} \dfrac{1}{\sqrt{\lambda_v\lambda_w}}H_u \\ \dfrac{1}{\sqrt{\lambda_u\lambda_w}}H_v \\ \dfrac{1}{\sqrt{\lambda_u\lambda_v}}H_w \end{pmatrix} \Rightarrow \tag{2.2.16a}$$

$$\begin{pmatrix} \dfrac{\partial}{\partial v}\dfrac{E_w}{\underline{\lambda}_u} - \dfrac{\partial}{\partial w}\dfrac{E_v}{\underline{\lambda}_u} \\ \dfrac{\partial}{\partial w}\dfrac{E_u}{\underline{\lambda}_v} - \dfrac{\partial}{\partial u}\dfrac{E_w}{\underline{\lambda}_v} \\ \dfrac{\partial}{\partial u}\dfrac{E_v}{\underline{\lambda}_w} - \dfrac{\partial}{\partial v}\dfrac{E_u}{\underline{\lambda}_w} \end{pmatrix} = -j\omega\mu\mathbf{H} \Rightarrow \tag{2.2.16b}$$

$$\nabla \times \mathbf{E} = -j\omega\mu\underline{\mathbf{\Lambda}} \cdot \mathbf{H}. \tag{2.2.16c}$$

We obtain an equivalence between (2.2.12) and (2.2.14) by doing the same with the remaining equations.

By considering plane waves, the dispersion equation can be given as

$$\omega^2\mu\varepsilon = (\underline{k}_u\underline{g}_u)^2 + (\underline{k}_v\underline{g}_v)^2 + (\underline{k}_w\underline{g}_w)^2, \tag{2.2.17}$$

where $\underline{\mathbf{k}} = (\underline{k}_u, \underline{k}_v, \underline{k}_w)$ is the wave number and $\underline{\mathbf{g}} = (\underline{g}_u, \underline{g}_v, \underline{g}_w)$ is the diagonal of $\underline{\mathbf{\Lambda}}$.

Obviously, this equation is satisfied for plane waves with the following wave vector

$$\underline{\mathbf{k}} = \frac{\omega}{c}\begin{pmatrix} \underline{g}_u^{-1}\sin(\theta)\cos(\phi) \\ \underline{g}_v^{-1}\sin(\theta)\sin(\phi) \\ \underline{g}_w^{-1}\cos(\theta) \end{pmatrix} = \frac{\omega}{c}\begin{pmatrix} \underline{g}_u^{-1} \\ \underline{g}_v^{-1} \\ \underline{g}_w^{-1} \end{pmatrix} \cdot \mathbf{n}, \tag{2.2.18}$$

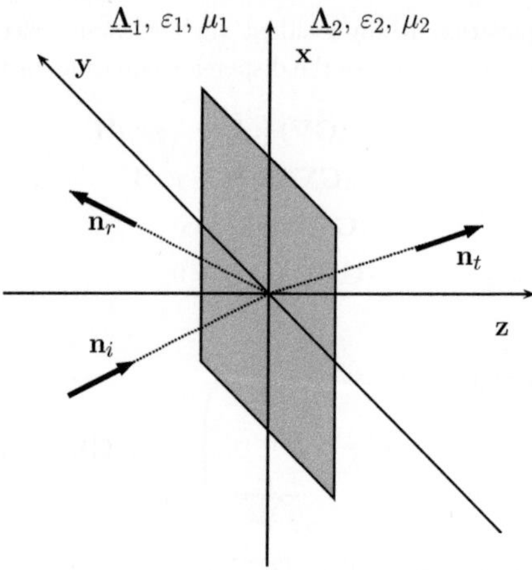

Figure 2.3: Transition between two planes of anisotropic materials

where **n** is the normal vector.

In the following the condition for plane wave propagation without reflection through the transition of two half planes with the interface plane $w = 0$ (Fig. 2.3) will be derived.

With the plane wave ansatz, reflection factors for TE[4]- and TM[5]-case are given as follows

$$R^{\text{TE}} = \frac{\underline{k}_{1w}\mu_2\underline{g}_{2w}^{-1} - \underline{k}_{2w}\mu_1\underline{g}_{1w}^{-1}}{\underline{k}_{1w}\mu_2\underline{g}_{2w}^{-1} + \underline{k}_{2w}\mu_1\underline{g}_{1w}^{-1}}, \tag{2.2.19}$$

$$R^{\text{TM}} = \frac{\underline{k}_{1w}\varepsilon_2\underline{g}_{2w}^{-1} - \underline{k}_{2w}\varepsilon_1\underline{g}_{1z}^{-1}}{\underline{k}_{1w}\varepsilon_2\underline{g}_{2w}^{-1} + \underline{k}_{2w}\varepsilon_1\underline{g}_{1z}^{-1}}. \tag{2.2.20}$$

Another condition is given by enforcing the wave numbers to match at the transition plane:

$$\omega\sqrt{\mu_1\varepsilon_1}\underline{g}_{1u}^{-1}\sin(\theta_e)\cos(\phi_e) = \omega\sqrt{\mu_2\varepsilon_2}\underline{g}_{2u}^{-1}\sin(\theta_e)\cos(\phi_e) \tag{2.2.21a}$$

$$\omega\sqrt{\mu_1\varepsilon_1}\underline{g}_{1v}^{-1}\sin(\theta_t)\sin(\phi_t) = \omega\sqrt{\mu_2\varepsilon_2}\underline{g}_{2v}^{-1}\sin(\theta_t)\sin(\phi_t). \tag{2.2.21b}$$

---

[4]transverse electric
[5]transverse magnetic

Equations (2.2.21a) and (2.2.21b) request the material coefficients ε, μ, $\underline{g}_u$, and $\underline{g}_v$ of the two media to coincide. From this, it can be proven in (2.2.19, 2.2.20) that no reflection occurs for all incident angles. The following material relations for the tensors $\underline{\mathbf{\Lambda}}_1$ and $\underline{\mathbf{\Lambda}}_2$ can then be derived from $\underline{g}_{1u} = \underline{g}_{2u}$ and $\underline{g}_{1v} = \underline{g}_{2v}$:

$$\frac{\underline{\lambda}_{1u}}{\underline{\lambda}_{2u}} = \frac{\underline{\lambda}_{1v}}{\lambda_{2v}} = \frac{\lambda_{2w}}{\lambda_{1w}}. \tag{2.2.22}$$

Typically, a frequency-dependent expression [25]

$$\underline{\lambda}_{2u} = \underline{\lambda}_{2v} = \underline{\lambda}_{2w}^{-1} = 1 + \frac{\sigma}{j\omega} \tag{2.2.23}$$

is used for the material properties of the PML medium because of its causal property [26].

Theoretically, the PML medium allows wave penetration without any reflection. However, this is not guaranteed in the discrete model as reflections may occur at boundaries of cells with high material gradient due to grid dispersion [27]. In order to reduce this effect, instead of an homogeneous material, several layers[6] with growing conductivity from layer to layer following an appropriate profile are used. For this purpose, power functions have been identified as more efficient as geometric ones [23]

$$\sigma(\omega) = \sigma_{max} \left(\frac{\omega}{\Delta\omega} \right)^q \text{ with } \sigma_{max} = -\frac{\varepsilon_0 c}{2\Delta\omega} \frac{q+1}{N_{\text{lay}}} \ln(R) \tag{2.2.24}$$

with the number of layers N_{lay}, the exponent q and the minimal reflection factor $R = \exp\left(\frac{2}{c} \int_0^\delta \sigma(\omega) d\omega \right)$ for a normal incident wave on a medium of penetration depth δ, c being the velocity of light, and $\Delta\omega$ the thickness of each PML layer. The structure of a typical PML medium with the above presented profile is shown in Fig 2.4.

Excitations

There are two main excitation types, which are waveguide and discrete ports. More details about waveguide ports which are not considered in this work can be found in [19, 27, 28]. The discrete port is defined along an edge with an assigned port impedance[7] on which a current or voltage is applied. They are appropriate for excitations within the computational domain.

The excitation occurs through current instead of voltage. The port is thus integrated by choosing an appropriate entry in the vector $\mathbf{\hat{J}}$ of (2.2.8b) and (2.1.2).

---

[6]Typically the number of layers varies from 4 to 8.

[7]The smallest element of the domain (edge) is taken in analogy to a dipole of infinitesimal small length.

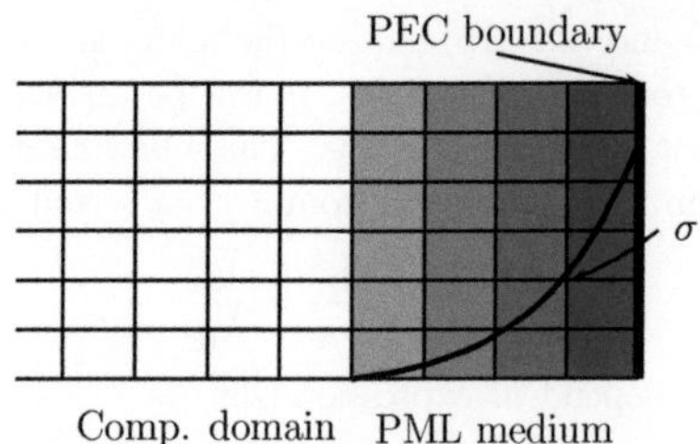

Figure 2.4: 4-layer PML medium with rising conductivity profile σ and electric boundary

2.2.4 Electrostatic

The constitutive equations of the electrostatic theory in FIT formalism are

$$\mathbf{C}\,\hat{\mathbf{e}} = 0 \tag{2.2.25a}$$

$$\widetilde{\mathbf{S}}\,\hat{\hat{\mathbf{d}}} = \mathbf{q}. \tag{2.2.25b}$$

With the well-known potential ansatz which fulfills (2.2.25a)

$$\vec{E} = -\mathrm{grad}\ \varphi \Longleftrightarrow \hat{\mathbf{e}} = \widetilde{\mathbf{S}}^T \boldsymbol{\Phi}, \tag{2.2.26}$$

(2.2.25b) can be derived to

$$\underbrace{\widetilde{\mathbf{S}}\mathbf{M}_\varepsilon \widetilde{\mathbf{S}}^T}_{\mathbf{F}} \boldsymbol{\Phi} = \mathbf{q} \tag{2.2.27}$$

which is the discrete equivalent of the Poisson equation. The matrix $\mathbf{F}$ is real, symmetric, and is of rank N_p, i.e. non singular .

2.3 FIT in System Representation

2.3.1 Curl Formulation

Combining the two first Maxwell's equations (2.1.3a and 2.1.3b) with the material properties (2.2.10) leads to the so-called curl-system [10]

$$\underbrace{\begin{pmatrix} \mathbf{M}_\varepsilon & 0 \\ 0 & \mathbf{M}_\mu^{-1} \end{pmatrix}}_{\mathbf{M}} \frac{\mathrm{d}}{\mathrm{d}t} \begin{pmatrix} \hat{\mathbf{e}} \\ \hat{\mathbf{h}} \end{pmatrix} = -\begin{pmatrix} \mathbf{M}_\sigma & -\widetilde{\mathbf{C}} \\ \mathbf{C} & 0 \end{pmatrix} \begin{pmatrix} \hat{\mathbf{e}} \\ \hat{\mathbf{h}} \end{pmatrix} + \begin{pmatrix} \hat{\hat{\mathbf{j}}} \\ 0 \end{pmatrix}. \tag{2.3.1}$$

This differential system is of first order and the number of degrees of freedom (DOFs) is $N = 6N_p = n_e + n_h$ (n_e and n_h are related to the electric and magnetic voltages, respectively).

Considering the dual relation in Section 2.2.1 ($\widetilde{\mathbf{C}} = \mathbf{C}^T$), introducing the normalized field strengths[8]

$$\widehat{\mathbf{e}}' = \mathbf{M}_\varepsilon^{1/2}\,\widehat{\mathbf{e}} \text{ with } \mathbf{M}_\varepsilon = \mathbf{M}_\varepsilon^{1/2}\mathbf{M}_\varepsilon^{1/2} \text{ and analogously } \widehat{\mathbf{h}}' = \mathbf{M}_\mu^{1/2}\,\widehat{\mathbf{h}}, \qquad (2.3.2)$$

and multiplying (2.3.1) with the inverse matrix $\mathbf{M}^{-1}$ leads to the system matrix $\mathbf{A}$ which is skew-symmetric. Assuming that the investigated structure is excited at m input ports, we can introduce a coupling matrix $\mathbf{R}$ with $\mathbf{M}_\varepsilon^{-1/2}\,\widehat{\mathbf{j}} = \mathbf{M}_\varepsilon^{-1/2}\mathbf{R}\mathbf{i} = \mathbf{R}'\mathbf{i}$ which relates the current ports, represented through the m-dimensional vector $\mathbf{i}$, to the current vector $\widehat{\mathbf{j}}$. This leads to the following system

$$\underbrace{\frac{\mathrm{d}}{\mathrm{d}t}\begin{pmatrix} \widehat{\mathbf{e}}' \\ \widehat{\mathbf{h}}' \end{pmatrix}}_{\dot{\mathbf{x}}} = -\underbrace{\begin{pmatrix} \mathbf{M}_\varepsilon^{-1/2}\mathbf{M}_\sigma\mathbf{M}_\varepsilon^{-1/2} & -\mathbf{M}_\varepsilon^{-1/2}\mathbf{C}^T\mathbf{M}_\mu^{-1/2} \\ \mathbf{M}_\mu^{-1/2}\mathbf{C}\mathbf{M}_\varepsilon^{-1/2} & 0 \end{pmatrix}}_{\mathbf{A}}\underbrace{\begin{pmatrix} \widehat{\mathbf{e}}' \\ \widehat{\mathbf{h}}' \end{pmatrix}}_{\mathbf{x}} + \underbrace{\begin{pmatrix} \mathbf{R}' \\ 0 \end{pmatrix}}_{\mathbf{B}}\mathbf{i}.$$

$$(2.3.3)$$

Analogously, an output coupling matrix can be defined which expresses the voltages $\mathbf{u}$ at the l output ports from the normalized voltage vector $\widehat{\mathbf{e}}'$

$$\mathbf{u} = \underbrace{\begin{pmatrix} \mathbf{L}\mathbf{M}_\varepsilon^{-1/2} & 0 \end{pmatrix}}_{\mathbf{C}}\begin{pmatrix} \widehat{\mathbf{e}}' \\ \widehat{\mathbf{h}}' \end{pmatrix}, \qquad (2.3.4)$$

where $\mathbf{L}$ is analogous to $\mathbf{R}$ in (2.3.3). Combining (2.3.3) and (2.3.4) leads to the well-known state-space representation

$$\dot{\mathbf{x}} = -\mathbf{A}\mathbf{x} + \mathbf{B}\mathbf{i} \qquad (2.3.5a)$$

$$\mathbf{u} = \mathbf{C}\mathbf{x} + \mathbf{D}\mathbf{i} \qquad (2.3.5b)$$

where $\mathbf{D} = 0$ in our case because there is no coupling between input and output. The matrix $\mathbf{A}$ is the $n \times n$ system matrix, and $\mathbf{x}$ is the n-dimensional state vector. In the current problem specification, the output and input ports are the same, so that the input and output coupling matrices have the relationship $\mathbf{C} = \mathbf{B}^T$ and $\mathbf{B}$ is of dimension $n \times m$ where m is the number of ports[9].

---

[8] As the material matrices are of diagonal form, their root can be obtained just by computing the root of their entries.

[9] $\mathbf{C}$ should here not be confounded with the curl operator.

As we assumed all signals to be harmonic, $\dot{\mathbf{x}}$ can be expressed as $j\omega\mathbf{x}$, which leads to the transformed state space representation

$$(j\omega\mathbf{I} + \mathbf{A})\mathbf{x} = \mathbf{Bi} \tag{2.3.6a}$$

$$\mathbf{u} = \mathbf{B}^T\mathbf{x}. \tag{2.3.6b}$$

In this system, the transfer function and the impedance function

$$\mathbf{Z}(j\omega) = \mathbf{B}^T(j\omega\mathbf{I} + \mathbf{A})^{-1}\mathbf{B}, \tag{2.3.7}$$

which are given by eliminating the state vector, are identical. The impedance function relates the currents and voltages at the different ports with each other as

$$Z_{[i,j]} = \frac{u_i}{i_j}\,\big|_{i_k=0\ \forall k\neq j}\,. \tag{2.3.8}$$

The curl system is of first order as it depends only in the first order on the frequency.

2.3.2 Curl-Curl Formulation

By combining the two first Maxwell's equations (2.1.3a and 2.1.3b) and thus eliminating either electric or magnetic field, we obtain the well known wave equation

$$\varepsilon\frac{\partial^2\vec{E}}{\partial t^2} + \sigma\frac{\partial\vec{E}}{\partial t} + \operatorname{curl}\mu^{-1}\operatorname{curl}\vec{E} = \frac{\partial\vec{J}}{\partial t}, \tag{2.3.9}$$

which is also called *curl-curl* system because of the two *curl*-operators. Analogously, by retaining $\hat{\mathbf{e}}$ as unknown vector, we obtain the discrete *curl-curl* state space representation

$$\mathbf{M}_\varepsilon\frac{\mathrm{d}^2}{\mathrm{d}t^2}\hat{\mathbf{e}} + \mathbf{M}_\sigma\frac{\mathrm{d}}{\mathrm{d}t}\hat{\mathbf{e}} + \underbrace{\mathbf{C}^T\mathbf{M}_{\mu^{-1}}\mathbf{C}}_{\mathbf{A}'_{CC}}\hat{\mathbf{e}} = \mathbf{Ri} \tag{2.3.10a}$$

$$\mathbf{u} = \mathbf{R}^T\hat{\mathbf{e}}. \tag{2.3.10b}$$

The elimination of $\hat{\mathbf{h}}$ leads to a system with $3N_p$ unknowns. By considering $\mathbf{C}^T = \mathbf{B} = \mathbf{M}_\varepsilon^{-1/2}\mathbf{R}$ and normalizing the state vector $\mathbf{x} = \mathbf{M}_\varepsilon^{1/2}\hat{\mathbf{e}}$, we obtain a symmetric system. With

$$\mathbf{A}_{CC} = \mathbf{M}_\varepsilon^{-1/2}\mathbf{A}'_{CC}\mathbf{M}_\varepsilon^{-1/2} \text{ and } \mathbf{K} = \mathbf{M}_\varepsilon^{-1/2}\mathbf{M}_\sigma\mathbf{M}_\varepsilon^{-1/2}, \tag{2.3.11}$$

the state space representation is given in the frequency domain as

$$\left((j\omega)^2\mathbf{I} + (j\omega)\mathbf{K} + \mathbf{A}_{CC}\right)\mathbf{x} = (j\omega)\mathbf{B}\mathbf{i}, \tag{2.3.12a}$$

$$\mathbf{u} = \mathbf{B}^T\mathbf{x}. \tag{2.3.12b}$$

The impedance function for *curl-curl* systems can then be expressed as

$$\mathbf{Z}(j\omega) = (j\omega)\mathbf{B}^T\left((j\omega)^2\mathbf{I} + (j\omega)\mathbf{K} + \mathbf{A}_{CC}\right)^{-1}\mathbf{B}. \tag{2.3.13}$$

This system is of second order with respect to the frequency term while it remains of first order in structures without losses ($\mathbf{K} = \mathbf{0}$) as the term ω^2 can be substituted e.g. by ω'. By introducing an additional variable, the Curl-Curl system can be transformed to a linear system. With $\mathbf{y} = \frac{q}{j\omega}\mathbf{x}$ (q being a given constant), the system (2.3.12) can be transformed into

$$j\omega\begin{pmatrix}\mathbf{x} \\ \mathbf{y}\end{pmatrix} = -\underbrace{\begin{pmatrix}\mathbf{K} & \frac{1}{q}\mathbf{A}_{CC} \\ -q\mathbf{I} & \mathbf{0}\end{pmatrix}}_{\mathbf{A}_l}\begin{pmatrix}\mathbf{x} \\ \mathbf{y}\end{pmatrix} + \begin{pmatrix}\mathbf{B} \\ \mathbf{0}\end{pmatrix}\mathbf{i}, \tag{2.3.14}$$

which has the same dynamic behavior as the system in (2.3.3). The matrices $\mathbf{A}_l$ and $\mathbf{A}$ have the same eigenvalues even though the symmetry properties are not kept. The constant q is essential in order to avoid a bad conditioned matrix. This goal is achieved with $q = \sqrt{\|\mathbf{A}_{CC}\|}$ as the blocks of $\mathbf{A}_l$ have then approximatively the same norm.

2.3.3 System with PML Absorber

PML Absorber of 1. Order

Without lost of generality, we assume the coordinate system $u - v - w$ with the surface normal of the interface along the w-axis. The material coefficients for the free space (material 1 according to Fig. 2.3) can be given as:

$$\varepsilon_1 = \varepsilon_0, \quad \mu_1 = \mu_0, \quad \underline{\mathbf{\Lambda}}_1 = \begin{pmatrix} 1 & 0 & 0 \\ 0 & 1 & 0 \\ 0 & 0 & 1 \end{pmatrix} \tag{2.3.15}$$

The w-dependency of the wave propagation in medium 2 is given as follows:

$$E(w) = E_0 \cdot e^{-jk_{2w}w} = E_0 \cdot e^{-j\frac{\omega}{c_0}\sqrt{\lambda_{2u}\lambda_{2v}}cos(\theta)} \tag{2.3.16}$$

In order to obtain an attenuation of the wave along the w-axis, we choose expression (2.2.23). This corresponds to a medium with electric and magnetic losses. From (2.2.22), we derive the material constant for the w-component as

$$\lambda_{2w} = \lambda_{2u}^{-1} = \lambda_{2v}^{-1} = \frac{1 - \frac{\sigma_w}{j\omega}}{1 + \left(\frac{\sigma_w}{j\omega}\right)^2}. \tag{2.3.17}$$

The negative conductivity indicates the active behavior of the PML on the w-axis. This non physical property confers the medium a perfect absorption behavior.

Starting from the discretized of the transformed Maxwell equations in (2.2.12),

$$\mathbf{C}\,\widehat{\mathbf{e}} = -j\omega\mathbf{M}_{\mu^{-1}}^{-1}\mathbf{\Lambda}\,\widehat{\mathbf{h}}, \tag{2.3.18a}$$

$$\mathbf{C}^T\,\widehat{\mathbf{h}} = j\omega\mathbf{M}_\varepsilon\mathbf{\Lambda}\,\widehat{\mathbf{e}} + \widehat{\widehat{\mathbf{j}}}_s, \tag{2.3.18b}$$

we can derive the implementation of the PML for FIT systems. In the following, we only consider (2.3.18a) as the derivation for (2.3.18b) is analogous,

$$\mathbf{C}\,\widehat{\mathbf{e}} = -j\omega\mathbf{M}_{\mu^{-1}}^{-1} \begin{pmatrix} 1 + \frac{\sigma_w}{j\omega} & 0 & 0 \\ 0 & 1 + \frac{\sigma_w}{j\omega} & 0 \\ 0 & 0 & \frac{1}{1+\frac{\sigma_w}{j\omega}} \end{pmatrix} \widehat{\mathbf{h}} \Rightarrow$$

$$\left(\mathbf{I} + \frac{1}{j\omega}\underbrace{\begin{pmatrix} 0 & 0 & 0 \\ 0 & 0 & 0 \\ 0 & 0 & \sigma_w \end{pmatrix}}_{\mathbf{M}_{\sigma 1}}\right)\mathbf{C}\,\widehat{\mathbf{e}} = -\mathbf{M}_{\mu^{-1}}^{-1}\left(j\omega\mathbf{I} + \underbrace{\begin{pmatrix} \sigma_w & 0 & 0 \\ 0 & \sigma_w & 0 \\ 0 & 0 & 0 \end{pmatrix}}_{\mathbf{M}_\sigma}\right)\widehat{\mathbf{h}}$$

The two Maxwell equations (2.3.18a and 2.3.18b) are then given as

$$\left(\mathbf{I} + \frac{1}{j\omega}\mathbf{M}_{\sigma 1}\right)\mathbf{C}\,\widehat{\mathbf{e}} = -\mathbf{M}_{\mu^{-1}}^{-1}\left(j\omega\mathbf{I} + \mathbf{M}_\sigma\right)\widehat{\mathbf{h}}, \tag{2.3.20a}$$

$$\left(\mathbf{I} + \frac{1}{j\omega}\mathbf{M}_{\sigma 1}\right)\mathbf{C}^T\,\widehat{\mathbf{h}} = \mathbf{M}_\varepsilon\left(j\omega\mathbf{I} + \mathbf{M}_\sigma\right)\widehat{\mathbf{e}} + \widehat{\widehat{\mathbf{j}}}_s. \tag{2.3.20b}$$

They can also be presented in matrix form which yields a second order system

$$((j\omega)^2\mathbf{I} + j\omega\mathbf{A}_1 + \mathbf{A}_0)\mathbf{x} = j\omega\mathbf{B}, \tag{2.3.21a}$$

$$\mathbf{u} = \mathbf{B}^T\mathbf{x}. \tag{2.3.21b}$$

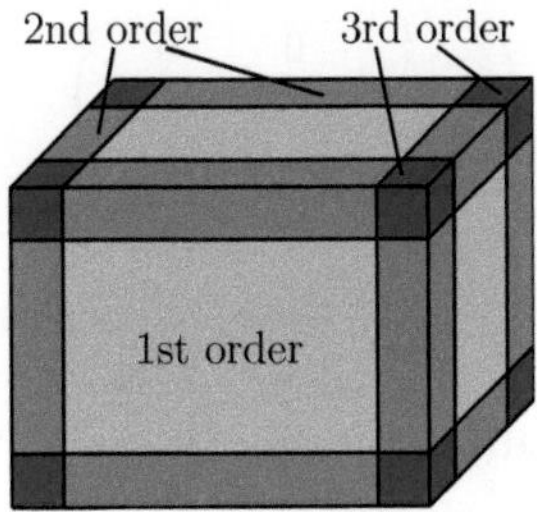

Figure 2.5: Higher order PML at the intersection of PML layers.

with the matrices

$$\mathbf{A}_0 = \begin{pmatrix} 0 & -\mathbf{M}_\varepsilon^{-1/2}\mathbf{M}_{\sigma 1}\mathbf{C}^T\mathbf{M}_{\mu^{-1}}^{1/2} \\ \mathbf{M}_{\mu^{-1}}^{1/2}\mathbf{M}_{\sigma 1}\mathbf{C}\mathbf{M}_\varepsilon^{-1/2} & 0 \end{pmatrix}, \quad \mathbf{B} = \begin{pmatrix} \mathbf{R} \\ 0 \end{pmatrix} \qquad (2.3.22a)$$

$$\mathbf{A}_1 = \begin{pmatrix} \mathbf{M}_\varepsilon^{-1/2}\mathbf{M}_\sigma\mathbf{M}_\varepsilon^{-1/2} & -\mathbf{M}_\varepsilon^{-1/2}\mathbf{C}^T\mathbf{M}_{\mu^{-1}}^{1/2} \\ \mathbf{M}_{\mu^{-1}}^{1/2}\mathbf{C}\mathbf{M}_\varepsilon^{-1/2} & \mathbf{M}_{\mu^{-1}}^{1/2}\mathbf{M}_\sigma\mathbf{M}_{\mu^{-1}}^{1/2} \end{pmatrix}, \quad \mathbf{x} = \begin{pmatrix} \mathbf{M}_\varepsilon^{-1/2}\,\hat{\mathbf{e}} \\ \mathbf{M}_{\mu^{-1}}^{1/2}\,\hat{\mathbf{h}} \end{pmatrix} \qquad (2.3.22b)$$

As already stated, the implemented PML is then appended to the computational domain and terminated either with a Perfect Magnetic Condition (PMC) or Perfect Electric Condition (PEC) boundary. By doing so, the depth of the absorbing material should be between 4 and 8 grid cells. This depth can be reduced to 2 by terminating with a first order Mur boundary condition [29]. An incident wave in the absorber medium is thus attenuated on its way to the termination and back after reflection before penetrating the computational domain. The reflection factor R_0 for a normal incident wave is a characteristic parameter of the PML boundary condition. The proper choice of this factor is described more detailed in [30].

PML Absorber of Higher Order

Stringing several PML boundary conditions together leads to new transition conditions and thus to new classes of absorbers. They are of order 1 to 3. PML media which are directly connected to the computational domain are of first order. Whereas absorbers of second and third order which are connected to the domain of computation through a grid point or an edge are the intersection of two PML media of lower order (Fig. 2.5).

The material properties of these absorbers can also be derived by requiring the transition planes to be reflectionless. Let us consider medium 2 being the intersection of two first order media ($1A$ and $1B$) with the following material tensors:

$$\boldsymbol{\Lambda}_{1A} = \begin{pmatrix} \underline{\omega}_{1Au}^{-1} & 0 & 0 \\ 0 & \underline{\omega}_{1Au} & 0 \\ 0 & 0 & \underline{\omega}_{1Au} \end{pmatrix} \tag{2.3.23a}$$

$$\boldsymbol{\Lambda}_{1B} = \begin{pmatrix} \underline{\omega}_{1Bw} & 0 & 0 \\ 0 & \underline{\omega}_{1Bw} & 0 \\ 0 & 0 & \underline{\omega}_{1Bw}^{-1} \end{pmatrix}, \quad \underline{\omega}_i = 1 + \frac{\sigma_i}{j\omega} \tag{2.3.23b}$$

The tensor of the second order absorber is given as follows [30]:

$$\boldsymbol{\Lambda}_2 = \begin{pmatrix} \frac{\underline{\omega}_{1Bw}}{\underline{\omega}_{1Au}} & 0 & 0 \\ 0 & \underline{\omega}_{1Au}\underline{\omega}_{1Bw} & 0 \\ 0 & 0 & \frac{\underline{\omega}_{1Au}}{\underline{\omega}_{1Bw}} \end{pmatrix}. \tag{2.3.24}$$

By doing the same, the material tensor of an third order absorber is derived as:

$$\boldsymbol{\Lambda}_3 = \begin{pmatrix} \frac{\underline{\omega}_v\underline{\omega}_w}{\underline{\omega}_u} & 0 & 0 \\ 0 & \frac{\underline{\omega}_u\underline{\omega}_w}{\underline{\omega}_v} & 0 \\ 0 & 0 & \frac{\underline{\omega}_u\underline{\omega}_v}{\underline{\omega}_w} \end{pmatrix} \tag{2.3.25}$$

The system equation with PML of higher order is the same as in (2.3.21), as only the matrices $\mathbf{A}_0$, $\mathbf{M}_{\sigma i}$ and $\mathbf{M}_{\sigma i}$ are changed

$$\mathbf{A}_0 = \begin{pmatrix} \mathbf{M}_\varepsilon^{-1/2}\mathbf{M}_{\sigma 2}\mathbf{M}_\varepsilon^{-1/2} & -\mathbf{M}_\varepsilon^{-1/2}\mathbf{M}_{\sigma 1}\mathbf{C}^T\mathbf{M}_{\mu^{-1}}^{1/2} \\ \mathbf{M}_{\mu^{-1}}^{1/2}\mathbf{M}_{\sigma 1}\mathbf{C}\mathbf{M}_\varepsilon^{-1/2} & \mathbf{M}_{\mu^{-1}}^{1/2}\mathbf{M}_{\sigma 2}\mathbf{M}_{\mu^{-1}}^{1/2} \end{pmatrix}, \tag{2.3.26}$$

the matrices $\mathbf{M}_{\sigma i}$ and $\mathbf{M}_{\sigma i}$ respectively for the second and third order are given as

$$\mathbf{M}_{\sigma 1}^{2.ord} = \begin{pmatrix} \sigma_u & 0 & 0 \\ 0 & 0 & 0 \\ 0 & 0 & \sigma_w \end{pmatrix}, \mathbf{M}_{\sigma 2}^{2.ord} = \begin{pmatrix} 0 & 0 & 0 \\ 0 & \sigma_u\sigma_w & 0 \\ 0 & 0 & 0 \end{pmatrix},$$

$$\mathbf{M}_{\sigma}^{2.ord} = \begin{pmatrix} \sigma_w & 0 & 0 \\ 0 & \sigma_u+\sigma_w & 0 \\ 0 & 0 & \sigma_u \end{pmatrix}, \tag{2.3.27}$$

$$\mathbf{M}_{\sigma 1}^{3.ord} = \begin{pmatrix} \sigma_u & 0 & 0 \\ 0 & \sigma_v & 0 \\ 0 & 0 & \sigma_w \end{pmatrix}, \mathbf{M}_{\sigma 2}^{3.ord} = \begin{pmatrix} \sigma_v\sigma_w & 0 & 0 \\ 0 & \sigma_u\sigma_w & 0 \\ 0 & 0 & \sigma_u\sigma_v \end{pmatrix}$$

$$\mathbf{M}_{\sigma}^{3.ord} = \begin{pmatrix} \sigma_v+\sigma_w & 0 & 0 \\ 0 & \sigma_u+\sigma_w & 0 \\ 0 & 0 & \sigma_u+\sigma_v \end{pmatrix}. \tag{2.3.28}$$

Again, this system can be linearized through the method introduced in (2.3.14). The system matrices are skew-symmetric as the $\hat{\mathbf{e}}$- and $\hat{\mathbf{h}}$-vectors are combined. Furthermore, the Curl-Curl ansatz is not recommended as merging of the equations (2.3.20a) and (2.3.20b) would lead to a considerably higher order system.

2.3.4 Scattering Parameters

Whereas the impedance matrices set the generalized port currents and voltages of a system in relation, the scattering parameters S are related to the wave amplitudes. S-parameters are of relevance in the high frequency context. This evolves from the fact that short or open conditions which are essential in determining impedances can not be properly realized in a measurement setup for microwave frequencies due to radiations and parasitic effects. Reflectionless terminations which are indispensable for the measurement of S-parameters can be set up very easily. Another advantage of S-parameters is the fact that they always exist while impedance matrices can not be computed for general ports.

The scattering matrix can be directly computed from the impedance matrix. First of all, the impedance matrix should be normalized with respect to the port-impedance matrix $\mathbf{Z}_0$:

$$\overline{\mathbf{Z}} = \mathbf{Z}_0^{-1/2}\mathbf{Z}\mathbf{Z}_0^{-1/2}. \tag{2.3.29}$$

The scattering matrix is then given as

$$\mathbf{S} = (\overline{\mathbf{Z}} - \mathbf{I})(\overline{\mathbf{Z}} + \mathbf{I})^{-1}. \tag{2.3.30}$$

Analogously, the impedance matrix can be computed from the scattering matrix as follows:

$$\mathbf{Z} = \mathbf{Z}_0^{1/2}(\mathbf{I} + \mathbf{S})(\mathbf{I} - \mathbf{S})\mathbf{Z}_0^{1/2}. \tag{2.3.31}$$

Some numerical errors occur by computing the impedance matrix using FIT. It is thus important to know how this errors are transmitted to the scattering parameters. In this respect, it has been shown in [10] for the one port case that the relative error in S is always smaller than in Z. Furthermore, practical tests attest that also for the multi-port case, the relative error of the scattering parameters is 0.5 to 2 orders lower than that of Z.

2.4 Eigenvalues of FIT Systems

By considering the *curl* and *curl-curl* (2.3.5, 2.3.12) systems without any excitation $(\mathbf{i} = \mathbf{0})$ and without losses in the *curl-curl* case, we obtain the eigenvalue problem in the form

$$\mathbf{A}\mathbf{x} = \lambda\mathbf{x}. \tag{2.4.1}$$

The solutions of this equation are the nontrivial *eigenvectors* $\mathbf{x}_i \neq \mathbf{0}$ and the *eigenvalues* λ_i.

2.4.1 Curl Formulation

As already stated, the curl system matrix $\mathbf{A}$ is skew-symmetric and real. It can be shown that such matrices have pairs of purely imaginary eigenvalues

$$\lambda_i^{\pm} = \mp j\omega, \tag{2.4.2}$$

where the number of $\lambda_i^{\pm}$ is half the dimension of the matrix $\mathbf{A}$. Furthermore, skew-matrices are diagonalizable.

2.4.2 Curl-Curl Formulation

The matrix of the *curl-curl* system is real symmetric and has, due to the following expression as product of two to each other transposed matrices

$$\begin{aligned}
\mathbf{A}_{CC} &= \mathbf{M}_{\varepsilon}^{-1/2}\mathbf{A}'_{CC}\mathbf{M}_{\varepsilon}^{-1/2} \\
&= (\mathbf{M}_{\varepsilon}^{-1/2}\mathbf{C}^{T}\mathbf{M}_{\mu^{-1}}^{1/2})(\mathbf{M}_{\varepsilon}^{-1/2}\mathbf{C}^{T}\mathbf{M}_{\mu^{-1}}^{1/2})^{T}
\end{aligned} \tag{2.4.3}$$

purely real and nonnegative eigenvalues

$$\lambda_i = \omega_i^2 \geq 0. \tag{2.4.4}$$

Furthermore, the eigenvectors are orthogonal. This confirms the physical property of lossless structures which have real eigenfrequencies and orthogonal modes.

Multiplying the eigenvalue equation of the Curl-Curl system with the normalized source operator $(\widetilde{\mathbf{S}}\mathbf{M}_{\varepsilon}^{1/2})$ leads to

$$\underbrace{\widetilde{\mathbf{S}}\widetilde{\mathbf{C}}}_{=\mathbf{0}}\mathbf{M}_{\mu^{-1}}\mathbf{C}\mathbf{M}_{\varepsilon}^{-1}\mathbf{x} = \omega^2\widetilde{\mathbf{S}}\mathbf{M}_{\varepsilon}^{1/2}\mathbf{x} = \omega^2\widetilde{\mathbf{S}}\widehat{\mathbf{d}} = \mathbf{0}. \tag{2.4.5}$$

This means that the eigenvalues or the divergence of the electric flux density of the corresponding eigenvectors vanish. They can thus be subdivided in the following categories [31]:

Static eigenvalues, $\mathbf{e}_s$: These modes which are called static because $\omega = 0$ can be represented as gradients of potentials ($\widehat{\mathbf{e}}_s = -(-\widetilde{\mathbf{S}}^{T})\phi$) and are thus irrotational[10]. On a grid with N_p nodes, without considering nodes with fixed

---

[10]rot grad = 0 ($\mathbf{C}\widetilde{\mathbf{S}}^{T} = \mathbf{0}$).

potential (PEC bodies or boundaries), i.e. only inner nodes, there are always less than N_p linear independent static modes.

Dynamic eigenvalues, e_d: These modes have a non vanishing curl part ($\omega \neq 0$) and are divergence free[11]. The Curl-Curl systems have $\sim 2N_p$ nonzero eigenvalues. They correspond to the modes which are propagable in closed structures.

Multiconductor eigenvalues, e_m: These modes are divergence free and irrotational ($\omega = 0$). They are related to charges on the surface of electric conducting regions. Assuming m from each other isolated conducting bodies in the computational domain, the number of multi-conductor eigenvalues is $m - 1$.

Any other, e_0: These modes do not have any physical interpretation and can be interpreted as artifacts. They are related either to ideal conducting edges or to components pointing outside of the computational domain and are also of static nature ($\omega = 0$). As they can be easily removed, they will not be considered in the following.

2.4.3 Regularization

The static eigenvalues do not play any role in the system behavior as only the dynamic and multiconductor modes are considered. However, they can cause tremendous mathematical difficulties while solving the eigenvalue problem. In fact, matrices containing such a huge kernel degrade the convergence performance of any eigensolver. Thus, a regularization becomes indispensable.

The so-called *Tree − Cotree* calibration [31] is one of the most efficient methods to remedy to that. Another one [18] is the grad-div calibration. It consists of adding a matrix $\mathbf{B}$ to the Curl-Curl operator

$$\mathbf{A}_{CC}\mathbf{x} = \omega^2\mathbf{x} \Rightarrow (\mathbf{A}_{CC} + \mathbf{B})\mathbf{x} = \gamma^2\mathbf{x}, \qquad (2.4.6)$$

so that the static eigenvalues can be shifted upwards without having any impact on the needed dynamic and multi-conductor modes

$$(\mathbf{A}_{CC} + \mathbf{B})\mathbf{x}_i = \gamma_i^2\mathbf{x}_i \wedge \widetilde{\mathbf{S}}\hat{\hat{\mathbf{d}}}_i = 0 \Rightarrow \gamma_i = \omega_i, \qquad (2.4.7a)$$

$$(\mathbf{A}_{CC} + \mathbf{B})\mathbf{x}_i = \gamma_i^2\mathbf{x}_i \wedge \widetilde{\mathbf{S}}\hat{\hat{\mathbf{d}}}_i \neq 0 \Rightarrow \gamma_i > 0. \qquad (2.4.7b)$$

In order to satisfy the above mentioned requirements, the kernel of the calibration matrix should be equal to the set of multi-conductor and dynamic modes. The grad-div

---

[11] div rot = 0.

operator achieves this condition as its kernel is divergence free and thus corresponds to the expected modes. So we choose

$$\mathbf{B} = \alpha \mathbf{M}_\varepsilon^{1/2} \widetilde{\mathbf{S}}^T \mathbf{D} \widetilde{\mathbf{S}} \mathbf{M}_\varepsilon^{1/2}. \tag{2.4.8}$$

Note that the grad-div operator has been symmetrized by considering the normalized fields ($\mathbf{x} = \mathbf{M}_\varepsilon^{1/2} \hat{\mathbf{e}}$). The regularized Curl-Curl operator is then similar to the *Nabla-square* operator

$$-\nabla^2 = \text{rot rot} - \text{grad div}. \tag{2.4.9}$$

Furthermore, the $\mathbf{P}_i$-matrices needed to construct the calibration matrix should be modified in such a way that no gradient is built for nodes in ideal conduction regions. As the modification on $\mathbf{P}_i$ would destroy the band structure and so the efficient assembling of $\mathbf{B}$, a different variant has been introduced. It consists on setting the appropriate entries of $\mathbf{D}$ to zero so that the above condition is satisfied.

The then displaced static eigenvalues land inside the spectrum and can be distinguished from the dynamic modes since they are divergence free. However they can worsen the efficiency of the solver if the minimum of the eigenvalues of $\mathbf{B}$ (without considering its kernel) is under the ground mode (smallest nonzero eigenvalue) and/or the maximum of the eigenvalues of $\mathbf{B}$ is above the maximum of the eigenvalues of $\mathbf{A}_{CC}$. The factor α can thus be adjusted in order to meet those constraints:

$$\alpha_{opt} = \frac{\max(\lambda_i(\mathbf{A}_{CC}))}{\max(\lambda_j(\mathbf{B}))}. \tag{2.4.10}$$

Alternatively, the grad-div term can be normalized with the matrix $\mathbf{D}_N$ in the following way [32]

$$\mathbf{B} = \alpha \mathbf{M}_\varepsilon^{1/2} \widetilde{\mathbf{S}}^T \underbrace{\widetilde{\mathbf{D}}_V^{-1} \widetilde{\mathbf{D}}_\nu \widetilde{\mathbf{D}}_\varepsilon^{-2}}_{\mathbf{D}_N} \widetilde{\mathbf{S}} \mathbf{M}_\varepsilon^{1/2}, \tag{2.4.11}$$

so that the relation between $\mathbf{A}_{CC}$ and $\mathbf{B}$ is set independent from material and geometry properties.

2.5 System Properties

The systems described in Section 2.3 are linear[12] and time invariant[13] (LTI). Thus, in the following, we will focus on LTI systems. Any bilinear transform defined as

$$F(s) = \int_0^\infty f(t)e^{-st}dt, \tag{2.5.1}$$

---

[12]The response to a linear combination of two inputs is the linear combination of the two corresponding outputs with the same coefficients.
[13]If $\mathbf{u}(t)$ is the response to $\mathbf{i}(t)$, then $\mathbf{u}(t - \tau)$ is the output for the delayed input $\mathbf{i}(t - \tau)$.

where $s = \sigma + j\omega$ can be characterized by its region of convergence (ROC) which represents the set of s values for which the integral in 2.5.1 converges absolutely.

2.5.1 Causality

The fundamental principle of causability evolves from the real world experience which states that an effect cannot precede its cause. Let us consider an LTI system defined by the following equation

$$y_i(t) = h_{[i,j]}(t) * x_j(t), \qquad (2.5.2)$$

where $y_i(t)$ and $x_j(t)$ are respectively the jth output and ith input signals, $h_{[i,j]}(t)$ the response and $*$ represents the convolution operation, then this system is causal if and only if all the elements of its response matrix $h_{i,j}(t)$ are vanishing for $t < 0$.

This condition is met if and only if the Laplace transform of the response matrix $\mathbf{H}(s)$

- is defined and analytic in a half-plane $Re\{s\} > \sigma_0$,

- grows not faster than a polynomial for $Re\{s\} > \sigma_0$,

where $\sigma_0 \in \mathbb{R}$. Maxwell's equations describe field propagations as causal systems and so does FIT. In the case of PML boundaries, the causality has been proved [10], as already stated, when using layers following (2.2.23).

2.5.2 Stability

The stability is an important property for electric systems. We distinguish two main definitions:

- A system is **transfer stable** or bounded-input,bounded-output (BIBO) stable if the output is bounded for all bounded inputs. The impulse response should satisfy the following condition:

$$\int_{-\infty}^{\infty} |h(t)|dt \leq M < \infty. \qquad (2.5.3)$$

- The **internal stability** is related to the system behavior at $t \rightarrow \infty$ for any arbitrary initial state $(\mathbf{x}_0(t))$ without excitation $(\dot{\mathbf{x}}(t) = \mathbf{A}\mathbf{x}(t))$. The system is asymptotical stable if the state variables decay to zero. It is called marginal stable if one or several states do not converge against zero but are bounded $(|x_i| \leq M_i < \infty)$.

In the Laplace domain, the conditions are given as follows [33]:

- A system is **asymptotical stable** if the eigenvalues of $\mathbf{A}$ have a positive real part.

- A system is **marginal stable** if the eigenvalues of $\mathbf{A}$ have a real part ≥ 0. Furthermore, the eigenvalues on the imaginary axis should be unique.

- A system is **transfer stable** if the eigenvalues of $\mathbf{A}$ which can not be reduced in the polynomial expression of $\mathbf{H}(s)$ have a positive real part. This coincides with the definition of asymptotical stable systems if no eigenvalue can be reduced.

FIT systems without losses in the curl-formulation are marginal stable as there eigenvalues are purely imaginary. However, they are not transfer stable, i.e. a harmonic excitation at a resonance frequency on the imaginary axis would lead to an unbounded output signal.

In the *curl-curl* formulation, systems without losses have eigenvalues $\lambda_i \geq 0$. Again, considering losses would shift the eigenvalues to the right half-plane, so that these systems are also marginal stable.

2.5.3 Passivity

A system is passive when it can not retrieve more energy than it has absorbed. In other words, it is unable to generate energy. The energy $w(t)$ should therefore satisfy the following condition for all t:

$$w(t) = \int_{-\infty}^{t} \mathbf{u}^T(\tau)\mathbf{i}(\tau)d\tau \geq 0. \tag{2.5.4}$$

System passivity is of great importance in transient simulations. As coupling of passive blocks also leads to a passive system, this property is the necessary condition to avoid numerical instabilities while combining different numerical methods (linear and nonlinear) in a simulation workflow. As most of the real structures under electromagnetical analysis are passive, it is important for the discrete model and furthermore, the macromodel of reduced order to keep this property.

The conditions for passivity of the impedance $\mathbf{Z}(s)$ in the Laplace domain are the following [34]:

- Any element of $\mathbf{Z}(s)$ is analytic for $Re\{s\} > 0$.

- $\mathbf{Z}(s^*) = \mathbf{Z}^*(s)$ for $Re\{s\} > 0$.

- $\mathbf{Z}^H(s) + \mathbf{Z}(s) \geq 0$ for $Re\{s\} > 0$.

Matrices which satisfy these three conditions are also called *positive real matrices* [35]. The following properties [36] of positive real matrices are of great importance for the passivity analysis of the FIT systems introduced above:

Theorem 1 *If $\mathbf{V}$ is a real constant $m \times n$-matrix and $\mathbf{G}(s)$ a positive real $m \times m$-matrix, then $\mathbf{V}^T\mathbf{G}(s)\mathbf{V}$ is also a positive real matrix.*

Theorem 2 *If $\mathbf{F}(s)$ and $\mathbf{G}(s)$ are positive real matrices, then $\mathbf{F}(s) + \mathbf{G}(s)$ is also positive real.*

Theorem 3 *If $\mathbf{G}(s)$ is positive real and $\mathbf{G}^H(s) + \mathbf{G}(s) > 0$ for $Re\{s\} > 0$, then $\mathbf{G}^{-1}(s)$ exists and is also positive real.*

As FIT systems can be represented in the form

$$\mathbf{Z}(s) = \mathbf{B}^T\left(\mathbf{Y}_1(s) + \mathbf{Y}_2(s) + \ldots\right)^{-1}\mathbf{B}, \tag{2.5.5}$$

passivity can be proved by checking if all $\mathbf{Y}_k$ are positive real. The first two theorems are typically fulfilled as only the third one should be analyzed. As already stated, the material matrices are diagonal and positive semi-definite in FIT systems, furthermore

$$\mathbf{A}_{CC}^H + \mathbf{A}_{CC} \geq 0, \quad \mathbf{A}_l^H + \mathbf{A}_l \geq 0 \tag{2.5.6}$$

because of their symmetry property, so that the analysis consists on checking for the passivity of the Laplace variable within the material definition. In the *curl-curl* case (2.3.12), we obtain

$$\mathbf{Y}^H(s) + \mathbf{Y} = \left(s\mathbf{I} + \mathbf{K} + \frac{1}{s}\mathbf{A}_{CC}\right)^H + \left(s\mathbf{I} + \mathbf{K} + \frac{1}{s}\mathbf{A}_{CC}\right) \tag{2.5.7a}$$

$$= 2\left(\sigma\mathbf{I} + \mathbf{K} + \frac{\sigma}{\sigma^2 + \omega^2}\mathbf{A}_{CC}\right) \geq 0 \text{ for } Re(s) > 0. \tag{2.5.7b}$$

The passivity for PML boundaries of first order can be proven analogously. By considering second order PML, we obtain a material of type $(1 + \sigma_u/s)(1 + \sigma_w/s)$ following (2.3.24). It follows for $\sigma \to 0$

$$\lambda_{uw}^* + \lambda_{uw} = 2 - \frac{2\sigma_u\sigma_w}{\omega^2}, \tag{2.5.8}$$

which means that the material is passive only for $\sigma_u\sigma_w < \omega^2$. As the values of the attenuation σ are of the order of ω, active components may occur. Thus, PML materials are generally non passive.

3 Model Order Reduction

Model order reduction methods have been first used in the field of control theory. It has been since then introduced to systems resulting from nodal analysis and discretization of Maxwell's equations. In this work, MOR is the main enabler to improve the efficiency of EMC simulation.

In this chapter, we present an overview of different MOR methods which can be typically divided in truncation and moment-matching methods. The methods of balanced and modal truncation as well as an explicit moment-matching method, AWE are briefly introduced. The main focus is set on implicit moment-matching methods related to the Lanczos' and Arnoldi's algorithms, with an emphasis on the passive preserving formulation which has been implemented in this work. Finally different methods to generate macromodels from the reduced systems are described.

3.1 Introduction

As discussed in Chapter 2, the discretization of the Maxwell's equations with FIT leads to a system of high dimension for complex structures. In order to compute the corresponding transfer function (Z- or S-parameters) in a wide frequency range, the systems of equations in (2.3.5 and 2.3.12) have typically to be solved at several frequency samples. This procedure is very expensive as the number of degrees of freedom is of order 10^6. On the other side, narrow-band resonances may not be detected if the frequency range is coarsely sampled. These narrow-band resonances would also represent a challenge for time domain computations as the slow energy dissipation lead to poor convergence behavior. MOR, already applied to solve Maxwell equations in [10–12] respectively for FIT, FEM and FVM (finite volume method), is a very robust method to remedy this problem.

Without lost of generality, we can describe the already introduced state space representation of FIT systems in the following way:

$$\begin{aligned} F(s)\mathbf{x} &= -\mathbf{A}\mathbf{x} + G(s)\mathbf{B}\mathbf{i} \\ \mathbf{u} &= \mathbf{C}\mathbf{x}, \end{aligned} \qquad (3.1.1)$$

where $F(s)$ and $G(s)$ are respectively equivalent to s and 1 in the *curl* case and to s^2 and $-s$ in the *curl-curl* case without losses as the cases with losses or PML will be treated separately. For simplicity, we will use the following notation

$$\Sigma = \left[\begin{array}{c|c} \mathbf{A} & \mathbf{B} \\ \hline \mathbf{C} & \end{array} \right] \in \mathbb{R}^{(n+m)\times(n+m)}, \tag{3.1.2}$$

MOR consists in approximating Σ with:

$$\Sigma_p = \left[\begin{array}{c|c} \mathbf{A}_p & \mathbf{B}_p \\ \hline \mathbf{C}_p & \end{array} \right] \in \mathbb{R}^{(p+m)\times(p+m)}, \tag{3.1.3}$$

where $p \ll n$ and the approximation error is small.

There are several MOR techniques which can be subdivided in two categories:

- Truncation methods

- Moment matching based methods

Truncation methods, among which balanced truncation [37–40] is the mostly used, provide error bounds and preserve stability, but are not appropriate for large scale systems as will be discussed in the following sections. In this work, we will focus on moment matching-based methods as they are numerically more efficient despite the fact that they have no global error bounds. In this scope, passivity preserving methods were implemented in order to guarantee stable circuit simulations involving nonlinear systems.

3.2 Truncation Methods

There are two sets of truncation methods which are suited for nonlinear (proper orthogonal decomposition, POD) and linear systems (Hankel-norm approximation, balanced truncation, singular perturbation) [41]. Before addressing the balanced reduction we will first introduce the so-called Hankel norm which plays an indispensable role for this class of MOR methods.

3.2.1 Hankel Norm

Given a matrix $\mathbf{A} \in \mathbb{R}^{n\times m}$, its singular value decomposition is defined as follows:

$$\mathbf{A} = \mathbf{U}_L \Sigma \mathbf{U}_R, \quad \Sigma = \mathrm{diag}(\sigma_1, \cdots, \sigma_n) \in \mathbb{R}^{n\times m}, \tag{3.2.1}$$

where $\sigma_1(\mathbf{A}) \geq \cdots \geq \sigma_n(\mathbf{A}) \geq 0$, are the singular values and $\sigma_1(\mathbf{A}) = \|\mathbf{A}\|_2$ is the 2-induced norm of $\mathbf{A}$. Moreover, the left and right singular vectors of $\mathbf{A}$ are orthonormal, $\mathbf{U}_R^T \mathbf{U}_R = \mathbf{I}$ and $\mathbf{U}_L^T \mathbf{U}_L = \mathbf{I}$. Assuming that $\sigma_k > 0$ and $\sigma_{k+1} = 0$, the rank of $\mathbf{A}$ is k and the *dyadic decomposition* of $\mathbf{A}$ is given as

$$\mathbf{A} = \sigma_1 \mathbf{u}_{L1} \mathbf{u}_{R1}^T + \cdots + \sigma_k \mathbf{u}_{Lk} \mathbf{u}_{Rk}^T. \tag{3.2.2}$$

The optimal approximation in the 2-norm is described in the following theorem [41]:

Theorem 4 *Provided that $\sigma_p > \sigma_{p+1}$, and $min_{rank\mathbf{X} \leq p}\|\mathbf{A} - \mathbf{X}\|_2 = \sigma_{p+1}$, a minimizer $\mathbf{X}_p$ is obtained by truncating the dyadic decomposition:* $\mathbf{X} = \sigma_1 \mathbf{u}_{L1} \mathbf{u}_{R1}^T + \cdots + \sigma_p \mathbf{u}_{Lp} \mathbf{u}_{Rp}^T.$

The Hankel operator[1] $\mathcal{H}$ of any system $\mathbf{\Sigma}$ built with the matrix $\mathbf{A}$ is known to be bounded and compact with the Hankel singular values satisfying the following inequalities

$$\sigma_1(\mathcal{H}) \geq \cdots \geq \sigma_n(\mathcal{H}) \geq 0, \tag{3.2.3}$$

with $\sigma_1 = \|\mathbf{\Sigma}\|_H$ being the *Hankel norm*. By the theorem stated above, any approximant $\hat{\mathcal{H}}$ of rank $p < n$ satisfies:

$$\|\mathcal{H} - \hat{\mathcal{H}}\|_2 \geq \sigma_{p+1}(\mathcal{H}). \tag{3.2.4}$$

The following holds especially for Hankel operators:

Theorem 5 *There exists a unique approximant $\mathcal{H}_p$ of rank p, which has Hankel structure and attains the lower bound: $\sigma_1(\mathcal{H} - \mathcal{H}_p) = \sigma_{p+1}(\mathcal{H}).$*

The Hankel singular values can be computed by solving the Lyapunov equations:

$$\mathbf{A}\mathbf{W}_O + \mathbf{W}_O\mathbf{A}^T + \mathbf{B}\mathbf{B}^T = 0 \tag{3.2.5a}$$

$$\mathbf{A}^T\mathbf{W}_C + \mathbf{W}_C\mathbf{A} + \mathbf{C}^T\mathbf{C} = 0, \tag{3.2.5b}$$

with $\mathbf{W}_O$ and $\mathbf{W}_C$ respectively being the observability and the controllability grammians. It can be shown that [9]:

$$\sigma_i(\mathbf{\Sigma}) = \sqrt{\lambda_i(\mathbf{W}_O\mathbf{W}_C)}, \tag{3.2.6}$$

and the error bound for optimal approximants is:

$$\sigma_{p+1} \leq \|\mathbf{\Sigma} - \mathbf{\Sigma}_p\|_\infty \leq 2(\sigma_{p+1} + \cdots + \sigma_n). \tag{3.2.7}$$

---

[1]This operator will be discussed explicitly in Section 3.3.

3.2.2 Balanced Truncation

The principle of balanced truncation is based on the fact that any system can be transformed to a basis where the states which are difficult to reach are simultaneously difficult to observe (*balanced system*). The reduced model is obtained by eliminating those states. The grammians of a balanced system have the following property:

$$\mathbf{W}_C = \mathbf{W}_O = \mathrm{diag}(\sigma_1, \cdots, \sigma_n). \tag{3.2.8}$$

A Cholesky decomposition of the system grammians $\mathbf{W}_C = \mathbf{X}\mathbf{X}^T$ and $\mathbf{W}_O = \mathbf{Y}\mathbf{Y}^T$ where $\mathbf{X}$ and $\mathbf{Y}$ are lower and their transposes upper triangular matrices and the following singular decomposition $\mathbf{X}^T\mathbf{Y} = \mathbf{U}_L\Sigma\mathbf{U}_R^T$ lead to the balancing transformation matrix

$$\mathbf{V}_{bt} = \mathbf{X}\mathbf{U}_L\Sigma^{-1/2} = (\Sigma^{-1/2}\mathbf{U}_R^T\mathbf{Y}^T)^{-1}. \tag{3.2.9}$$

Let us consider a balanced system with grammians equal to $\Sigma = diag(\Sigma_1, \Sigma_2)$, where $\Sigma_1 \in \mathbb{R}^{p \times p}$, and Σ_2 contains the small Hankel singular values. The truncation consists of projecting the original system onto the matrix obtained by eliminating the columns of $\mathbf{V}_{bt}$ corresponding to the small Hankel values:

$$\Sigma_p = \left[\begin{array}{c|c} \mathbf{A}_p & \mathbf{B}_p \\ \hline \mathbf{C}_p & \end{array} \right] \tag{3.2.10}$$

The main advantages of the balanced truncation as already stated are the preservation of stability and the global error bound. However, the resolution of the Lyapunov equations and the singular value decomposition which are of order $\mathcal{O}(n^3)$ makes this method inefficient for large-scale problems.

3.2.3 Modal Truncation

Whereas the methods presented in Sections 3.2.1 and 3.2.2 are based on the singular decomposition, modal truncation relies on the eigenvalue decomposition of a system. By applying it to $\mathbf{A}_{CC} = \mathbf{V}\Lambda\mathbf{V}^T$ ($\mathbf{V}^{-1} = \mathbf{V}^T$ as $\mathbf{A}_{CC}$ is the symmetric matrix resulting from the curl-curl equation), the transfer function in the curl-curl case without losses is derived to

$$\mathbf{H}(j\omega) = \mathbf{B}^T \left((j\omega)^2\mathbf{I} + \mathbf{V}\Lambda\mathbf{V}^T \right)^{-1} \mathbf{B} \tag{3.2.11a}$$

$$= \mathbf{B}^T\mathbf{V} \left((j\omega)^2\mathbf{I} + \Lambda \right)^{-1} \mathbf{V}^T\mathbf{B}. \tag{3.2.11b}$$

With the eigenvectors $\mathbf{V} = [\underbrace{\mathbf{v}_1, \mathbf{v}_2, \ldots, \mathbf{v}_p}_{\mathbf{V}_1}, \underbrace{\mathbf{v}_{p+1}, \ldots, \mathbf{v}_N}_{\mathbf{V}_2}]$ and $\lambda_i = \omega_i$ (2.4.4), it follows

$$\mathbf{Z}(j\omega) = j\omega \sum_{i=1}^{N} \frac{\mathbf{B}^T \mathbf{v}_i \mathbf{v}_i^T \mathbf{B}}{\omega_i^2 - \omega^2} \tag{3.2.12a}$$

$$= j\omega \sum_{i=1}^{p} \frac{\mathbf{B}^T \mathbf{v}_i \mathbf{v}_i^T \mathbf{B}}{\omega_i^2 - \omega^2} + \underbrace{j\omega \sum_{i=p+1}^{N} \frac{\mathbf{B}^T \mathbf{v}_i \mathbf{v}_i^T \mathbf{B}}{\omega_i^2 - \omega^2}}_{\mathbf{Z}_{\text{corr}}} \tag{3.2.12b}$$

The modal truncation consists thus of considering the first part of the sum in (3.2.12b) and an approximation $\widetilde{\mathbf{Z}}_{\text{corr}}$ of the correction term $\mathbf{Z}_{\text{corr}}$

$$\mathbf{Z}_r(j\omega) = j\omega \sum_{i=1}^{p} \frac{\mathbf{B}^T \mathbf{v}_i \mathbf{v}_i^T \mathbf{B}}{\omega_i^2 - \omega^2} + \widetilde{\mathbf{Z}}_{\text{corr}}(j\omega). \tag{3.2.13}$$

The state space representation of the first part is given as

$$(\mathbf{V}_1^T \mathbf{A} \mathbf{V}_1 + s\mathbf{I})\mathbf{x}_1 = \mathbf{V}_1^T \mathbf{B} i \tag{3.2.14a}$$

$$\mathbf{u}_1 = \mathbf{B}^T \mathbf{V}_1 \mathbf{x}_1, \tag{3.2.14b}$$

where $\mathbf{x}_1 \in \mathbb{R}^p$ and $\mathbf{u}_1 \in \mathbb{R}^m$ are the state and output variables of the truncated system without the correction term. A complex derivation of the correction term has been presented in [42]. By solving for the complementary system, an approximation of the admittance $\mathbf{Y}_p(j\omega) = \mathbf{Z}_p^{-1}(j\omega)$ is computed. The relation between poles and zeros of admittance and impedance then yields a good approximation of $\mathbf{Z}_{\text{corr}}(s)$. A more simple derivation presented in [31] is based on the exact solution at one or more frequency samples (ω_k). For the modes far away from the concerned frequency range, the following holds

$$\frac{\omega_k}{\omega_i^2 - \omega_k^2} \approx \frac{\omega_k}{\omega_i^2} \text{ for } |\omega_k| \ll |\omega_i|. \tag{3.2.15}$$

The contribution of the not considered modes can then be derived as a linear function of the exact correction terms at the chosen samples

$$\widetilde{\mathbf{Z}}_{\text{corr}}(j\omega) = \frac{\omega}{\sum_k \omega_k} \sum_k \mathbf{Z}_{\text{corr}}(j\omega_k). \tag{3.2.16}$$

The dominant eigenvectors[2] are not only contained in the frequency range but may reside far away. This enhances the number of eigenvalues to be computed and since

---

[2]The dominant eigenvectors can be determined by comparing the quotient of the terms in 3.2.13. Thus, eigenvectors which are almost parallel to the input vectors may be dominant even though they are far away from the considered frequency interval.

the eigenvalue computation is very time consuming [43], the efficiency of the modal truncation is tremendously degraded for large scale problems.

3.3 Asymptotic Waveform Evaluation

One of the most popular moment matching based methods for electromagnetic fields is the asymptotic waveform evaluation (AWE) [44] which is also called explicit moment matching method. For simplification, we consider the following *single-input single-output* (SISO) system:

$$(s\mathbf{I} + \mathbf{A})\mathbf{x} = \mathbf{b}i \tag{3.3.1a}$$

$$u = \mathbf{c}\mathbf{x}, \tag{3.3.1b}$$

where $\mathbf{A} \in \mathbb{R}^{n \times n}$ and $\mathbf{x}$, $\mathbf{b}$, $\mathbf{c} \in \mathbb{R}^n$. This representation is equivalent to the curl system (2.3.5) and its transfer function

$$H(s) = \mathbf{c}(s\mathbf{I} + \mathbf{A})^{-1}\mathbf{b} \tag{3.3.2}$$

is proportional to the curl-curl impedance function (2.3.12) without losses[3].
Assuming the following variable transformation

$$\hat{s} = s - s_0, \tag{3.3.3}$$

the transfer function (3.3.2) can be derived to

$$H(\hat{s}) = \mathbf{c} \left(\mathbf{I} + \hat{s}\hat{\mathbf{A}}\right)^{-1} \hat{\mathbf{b}}, \tag{3.3.4}$$

with

$$\hat{\mathbf{A}} = (\mathbf{A} + s_0\mathbf{I})^{-1} \text{ and } \hat{\mathbf{b}} = \hat{\mathbf{A}}\mathbf{b}. \tag{3.3.5}$$

With the vectors $\mathbf{x}_k$ obtained from the following iteration

$$\mathbf{x}_k = \hat{\mathbf{A}}^k \hat{\mathbf{b}}, \tag{3.3.6}$$

the transfer function can be expressed in geometric series

$$H(\hat{s}) = \sum_{k=0}^{\infty} \underbrace{\mathbf{c}\mathbf{x}_k}_{m_k} \hat{s}^k, \tag{3.3.7}$$

---

[3]The frequency factor can be obviously changed from s to s^2 without loss of generality.

where the moments m_k are equivalent to Taylor coefficients with the expansion frequency s_0.

AWE consists in approximating $H(\hat{s})$ with a transfer function of reduced order $H_p(\hat{s})$, so that the first $2p$ ($p \ll n$) moments of $H(\hat{s})$ and $H_p(\hat{s})$ are equivalent at the expansion frequency [44]

$$H(\hat{s}) \approx H_p(\hat{s}) = \sum_{k=0}^{2p-1} m_k \hat{s}^k. \qquad (3.3.8)$$

Taylor series have a convergence radius which is bounded by the poles of the function, so that this approximation is available only for a very small frequency range. In other to remedy this constraint, the Taylor series have been combined with the *Padé approximation* which is not limited by any pole.

By considering the eigenvalue decomposition of $\hat{\mathbf{A}} = \mathbf{V}\mathbf{\Lambda}\mathbf{V}^{-1}$ in (3.3.4), we obtain

$$H(\hat{s}) = \underbrace{\mathbf{c}\mathbf{V}}_{\mathbf{l}^T} \left(\mathbf{I} + \hat{s}\mathbf{\Lambda}\right)^{-1} \underbrace{\mathbf{V}^{-1}\hat{\mathbf{b}}}_{\mathbf{r}}. \qquad (3.3.9)$$

This expression can easily be transformed in a rational polynomial[4]

$$H(\hat{s}) - \sum_{k=1}^{n} \frac{l_k r_k}{1 + \hat{s}\lambda_k} = \frac{b_{n-1}\hat{s}^{n-1} + \cdots + b_1\hat{s} + b_0}{a_n\hat{s}^n + \cdots + a_1\hat{s} + 1}. \qquad (3.3.10)$$

The Padé approximation consists in truncating the original system with a polynomial of reduced order

$$H(\hat{s}) \approx H_p(\hat{s}) = \frac{b_{p-1}\hat{s}^{p-1} + \cdots + b_1\hat{s} + b_0}{a_p\hat{s}^p + \cdots + a_1\hat{s} + 1}. \qquad (3.3.11)$$

By setting (3.3.8) and (3.3.4) equal, we obtain a Padé approximation with the first $2p$ moments being identic to those of the original system. The coefficients $a_1, \cdots, a_p$ are computed by solving the following system of equations:

$$\underbrace{\begin{pmatrix} m_0 & m_1 & \cdots & m_{p-1} \\ m_1 & m_2 & \cdots & m_p \\ \vdots & \vdots & \ddots & \vdots \\ m_{p-1} & m_p & \cdots & m_{2p-2} \end{pmatrix}}_{\mathbf{M}_p} \begin{pmatrix} a_p \\ a_{p-1} \\ \vdots \\ a_1 \end{pmatrix} = - \begin{pmatrix} m_p \\ m_{p+1} \\ \vdots \\ m_{2p-1} \end{pmatrix}. \qquad (3.3.12)$$

---

[4]Note that the order of the numerator polynomial is always smaller than that of the denominator by one.

$\mathbf{M}_p$ is the so-called Hankel operator resulting from the system (3.3.1b). The remaining coefficients $b_0, \cdots, b_{p-1}$ can be obtained with the following recursion

$$
\begin{aligned}
b_0 &= m_0 \\
b_1 &= m_1 + b_1 m_0 \\
&\vdots \\
b_{p-1} &= m_{p-1} + \sum_{i=1}^{p-1} b_i m_{p-i-1}.
\end{aligned}
\tag{3.3.13}
$$

This method can be also applied to systems which are polynomial in s. Curl-curl systems with losses or systems with PML boundaries can be expressed as

$$
(s^2\mathbf{I} + s\mathbf{A}_1 + \mathbf{A}_0)\mathbf{x} = s\mathbf{bi}
\tag{3.3.14a}
$$

$$
\mathbf{u} = \mathbf{cx}.
\tag{3.3.14b}
$$

Again, by substituting s through $\hat{s}$, we obtain:

$$
H(\hat{s}) = \mathbf{c}(\hat{\mathbf{A}}_0 + \hat{s}\hat{\mathbf{A}}_1 + \hat{s}^2\hat{\mathbf{A}}_2)^{-1}\hat{s}\mathbf{b}_0,
\tag{3.3.15}
$$

with

$$
\hat{\mathbf{A}}_0 = \mathbf{I} + s_0\mathbf{A}_0 + s_0^2\mathbf{A}_1
\tag{3.3.16a}
$$

$$
\hat{\mathbf{A}}_1 = 2s_0\mathbf{I} + \mathbf{A}_1
\tag{3.3.16b}
$$

$$
\hat{\mathbf{A}}_2 = \mathbf{I}
\tag{3.3.16c}
$$

The moments of the system are then given as

$$
m_i = \mathbf{cx}_i,
\tag{3.3.17}
$$

where the $\mathbf{x}_i$ are computed with the recursion

$$
\begin{aligned}
\mathbf{x}_0 &= \hat{\mathbf{A}}_0^{-1}\hat{\mathbf{b}}_0 \\
\mathbf{x}_1 &= -\hat{\mathbf{A}}_0^{-1}\hat{\mathbf{A}}_1\mathbf{x}_0 \\
\mathbf{x}_2 &= \hat{\mathbf{A}}_0^{-1}(-\hat{\mathbf{A}}_1\mathbf{x}_1 - \hat{\mathbf{A}}_2\mathbf{x}_0) \\
&\vdots \\
\mathbf{x}_p &= \hat{\mathbf{A}}_0^{-1}(-\hat{\mathbf{A}}_1\mathbf{x}_{p-1} - \hat{\mathbf{A}}_2\mathbf{x}_{p-2}).
\end{aligned}
\tag{3.3.18}
$$

Contrary to the classical AWE, no Padé approximation is performed to compute the reduced system, but the following projection

$$
H_p = \mathbf{c}\mathbf{V}_p \left(\sum_{k=0}^{2} (s-s_0)^k \mathbf{V}_p^T \hat{\mathbf{A}}_k \mathbf{V}_p \right)^{-1} \mathbf{V}_p^T \hat{\mathbf{b}},
\tag{3.3.19}
$$

where $\mathbf{V}_p$ is an orthonormal basis of the space resulting from the iteration (3.3.19), obtained through QR decomposition

$$\widetilde{\mathbf{V}}_p = [\tilde{\mathbf{x}}_0, \tilde{\mathbf{x}}_1, \ldots, \tilde{\mathbf{x}}_p] = \mathbf{V}_p \mathbf{U}. \tag{3.3.20}$$

The first p moments of reduced and original systems match together which means that this reduction process is a Padé-like approximation.

Unfortunately, the convergence rate of this method is satisfying just for values of p up to 10. Whereas it stagnates for higher values of p. This is due to the fact that the vectors $\mathbf{x}_i$ get more and more linear dependent as they tend to the dominant eigenvector of $\hat{\mathbf{A}}$ or $\hat{\mathbf{A}}_0^{-1}\hat{\mathbf{A}}_1$ (Mises iteration) [9]. The linear dependence of those vectors in turn leads to an ill-conditioned matrix $\mathbf{M}_p$ and thus worsens the efficiency of the method.

In order to mitigate this instability, several methods have been proposed such as the complex frequency hopping (CFH) which consists of approximating the systems with reduced models at several expansion frequencies [45, 46]. A more efficient way is to match the moments implicitly instead of computing them explicitly by means of Krylov methods which increase the numerical stability.

3.4 Implicit Moment Matching Methods

3.4.1 Krylov Space

As already mentioned, Krylov subspaces play a central role in robust MOR techniques. The Krylov space resulting from the matrix $\mathbf{A}$ and the start vector $\mathbf{b}$ is given as [47]

$$\mathcal{K}_p(\mathbf{A}, \mathbf{b}) = \mathrm{span}\{\mathbf{b}, \mathbf{A}\mathbf{b}, \cdots, \mathbf{A}^{p-1}\mathbf{b}\}. \tag{3.4.1}$$

The rank of $\mathcal{K}_p$ is smaller than the rank of $\mathbf{A}$ and Krylov spaces are invariant to matrix scales and shifts:

$$\mathcal{K}_p(t\mathbf{A} + s\mathbf{I}, \mathbf{b}) = \mathcal{K}_p(\mathbf{A}, \mathbf{b}). \tag{3.4.2}$$

This definition can be easily extended to MIMO (multiple-input multiple-output) systems with several start vectors. The so-called Block-Krylov space with $\mathbf{B} = [\mathbf{b}_1, \cdots, \mathbf{b}_m]$ is given as follows

$$\mathcal{K}_p(\mathbf{A}, \mathbf{B}) = \mathrm{span}\{\mathbf{B}, \mathbf{A}\mathbf{B}, \cdots, \mathbf{A}^{p-1}\mathbf{B}\}. \tag{3.4.3}$$

Obviously, this space is equivalent to the vector basis introduced in the AWE iteration (3.3.6). However, in combination with Lanczos and Arnoldi [48, 49] algorithms, originally introduced to solve linear equation systems or eigenvalue problems, the instability of this iteration can be efficiently mitigated. As the focus is on passivity preserving methods, a passive formulation has been implemented. As usual Krylov spaces cannot consider systems which are polynomial in s, a stable extension of the iteration (3.3.19) will be considered.

3.4.2 Lanczos and Arnoldi Algorithms

These algorithms have been originally introduced for building an orthogonal basis of the Krylov space $\mathcal{K}_p$ for the solution of large scale equation systems [47] or eigenvalue problems [43]. They have been introduced in 1994 for model order reduction purposes [50]. Arnoldi's method aims to reduce a dense generally non-hermitian matrix into Hessenberg form. One variant of this algorithms with the modified Gram-Schmidt method for efficient orthogonalization is given in Algorithm 3.4.1.

Algorithm 3.4.1 Arnoldi's algorithm

$\mathbf{v}_1 = \mathbf{b}/\|\mathbf{b}\|_2$
for $j = 1$ to $p - 1$ **do**
  $\mathbf{w}_j = \mathbf{A}\mathbf{v}_j$
  for $i = 1$ to j **do**
    $h_{i,j} = (\mathbf{w}_j, \mathbf{v}_i)$
    $\mathbf{w}_j = \mathbf{w}_j - h_{i,j}\mathbf{v}_i$
  end for
  $h_{j+1,j} = \|\mathbf{w}_j\|_2$
  if $h_{j+1,j} = 0$ **then**
    Stop
  end if
  $\mathbf{w}_j = \mathbf{w}_j/h_{j+1,j}$
end for

This procedure can be easily extended to the block-case. An important property of the Arnoldi's algorithm is illustrated for the general block-iteration in Fig. 3.1. The matrix $\mathbf{H}_p$ is of Hessenberg form and the term $\mathbf{W}_p$ is proportional to the residuum in the scope of linear equations.

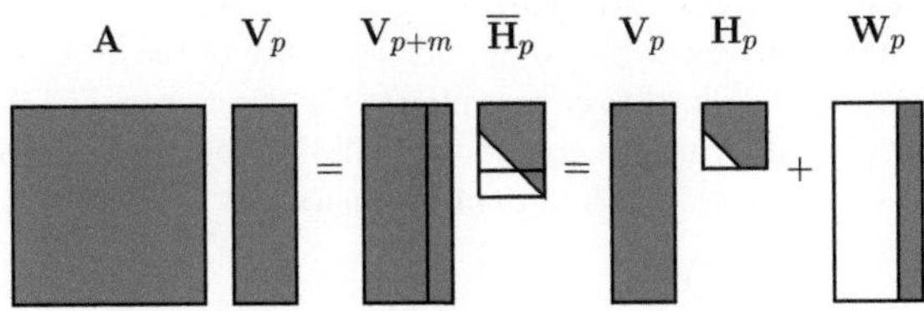

Figure 3.1: Arnoldi property.

For symmetric matrices $\mathbf{A}$, the Hessenberg matrix becomes tridiagonal[5] and the Arnoldi's algorithm can be substituted with the Lanczos' iteration given in Algorithm 3.4.2 for SISO systems.

Algorithm 3.4.2 Lanczos' algorithm

$\mathbf{v}_1 = \mathbf{b}/\|\mathbf{b}\|_2$
for $j = 1$ **to** $p - 1$ **do**
$\quad \mathbf{w}_j = \mathbf{A}\mathbf{v}_j$
$\quad i_0 = max(1, j - 2)$
$\quad$**for** $i = i_0$ **to** j **do**
$\quad\quad h_{i,j} = (\mathbf{w}_j, \mathbf{v}_i)$
$\quad\quad \mathbf{w}_j = \mathbf{w}_j - h_{i,j}\mathbf{v}_i$
$\quad$**end for**
$\quad h_{j+1,j} = \|\mathbf{w}_j\|_2$
$\quad$**if** $h_{j+1,j} = 0$ **then**
$\quad\quad$Stop
$\quad$**end if**
$\quad \mathbf{w}_j = \mathbf{w}_j/h_{j+1,j}$
end for

This iteration guarantees in exact arithmetic that the vectors $\mathbf{v}_i$ are orthogonal. The main advantage of this method is the reduced complexity in computation and storage of the orthogonalization process. In order to make this advantage available to

---

[5]This corresponds to a band-matrix with 2m secondary diagonals in the m-block case.

unsymmetric matrices, the Bi-Lanczos algorithm can be applied instead of Arnoldi. However, it has been shown that exact orthogonality of the vectors is lost at some point [47]. Despite the efforts to either mitigate this effect or recover the orthogonality of the vectors [51], the Arnoldi's algorithm is still the most robust method to generate orthogonal Krylov spaces, also for symmetric matrices.

Projection

Starting from the shifted transfer function (3.3.4), the reduced model is obtained by projecting the original system onto the orthogonal basis $\mathbf{V}_p$ and $\mathbf{W}_p$ ($\mathbf{W}_p^T \mathbf{V}_p = \mathbf{I}_p$) resulting respectively from the Krylov spaces $\mathcal{K}_p(\hat{\mathbf{A}}, \hat{\mathbf{b}})$ and $\mathcal{K}_p(\hat{\mathbf{A}}^T, \mathbf{c})$:

$$H_p = \mathbf{c}\mathbf{V}_p \left(\mathbf{W}_p^T \mathbf{V}_p + (s - s_0)\mathbf{W}_p^T \hat{\mathbf{A}} \mathbf{V}_p \right)^{-1} \mathbf{W}_p^T \hat{\mathbf{b}} \tag{3.4.4a}$$

$$= \mathbf{c}_p (\mathbf{I} + (s - s_0)\mathbf{T}_p)^{-1} \mathbf{b}_p. \tag{3.4.4b}$$

This projection method, also called Padé via Lanczos (PVL), was first introduced in [50] for network analysis purposes. It can easily be proven that the first $2p$ moments of the reduced and original systems are identical:

$$\mathbf{c}(-\hat{\mathbf{A}})^i \hat{\mathbf{b}} = \mathbf{c}(-\mathbf{T}_p)^i \hat{\mathbf{b}}. \tag{3.4.5}$$

The reduced system is thus a Padé approximation with moment matching. This can also be extended to MIMO systems where the moments are matched block-wise[6].

A very important issue is the choice of the interpolation point s_0. This has been intensively studied in [52]. The reduced systems with imaginary interpolations approximate first the poles nearest to s_0 whereas the approximation of distant poles needs more iterations. Their main advantage is thus the good local approximation of the transfer function in the neighborhood of s_0. The best choice for the interpolation point is at the center of gravity of the poles appearing in the frequency interval of interest. As this information is typically not available a priori, the middle of the frequency interval represents a good choice.

Real interpolation points allow a fast convergence of poles in the magnitude of s_0 and enables a broader but coarser convergence rate. In fact, numerous iterations are required to find sharp resonances (sharply damped poles) [52].

---

[6]The number of matching moments, $2m_p$ ($m_p = \text{floor}(q/m)$) depends on the number of ports m.

Multi-point Padé

The multi-point Padé approximation, also called rational Krylov [52–55], has been introduced in order to improve the convergence rate of the method. It consists of considering several interpolation points for the Padé approximation. The projection matrix $\mathbf{V}_p$ is then given as the orthogonal basis of the union of the Krylov subspaces at the considered k interpolation points:

$$\mathbf{V}_p \subset \bigcup_{k=1}^{K} \mathcal{K}_{p_k}((\mathbf{A} + s_{\mathbf{k}}\mathbf{I})^{-1}, (\mathbf{A} + s_k\mathbf{I})^{-1}\mathbf{B}), \qquad (3.4.6)$$

$$\text{with } \mathbf{V}_p^T \mathbf{V}_p = \mathbf{I} \text{ and } p = \sum_{k=1}^{K} p_k.$$

The dimension p_k can be set to 1 for every s_k or adjusted according to the relevance of the interpolation points. It is obvious that this technique is more time consuming than the single-point variant as the original matrix $\mathbf{A}$ has to be inverted K times. The main issue for an optimal trade-off between computational complexity and convergence acceleration is thus the choice of number and location of the interpolation points.

There are two strategies which have been introduced in order to address this point:

- The first consists of setting a priori the interpolation points s_k and building accordingly the projection matrix. This method offers a linear scalability for parallelization as the different processes are independent from each other. However, the location of the interpolation points is empiric and the method does not always guarantee a better efficiency as it leads to numerical problems as reported in [53].

- The second strategy aims for choosing adaptively the interpolation points after each p_k iterations. An empirical method for adaptive choice as the bisection search would suffer from the same drawback as the above mentioned strategy. A more efficient technique is to choose the next expansion point in the region with the highest approximation error. It is worth to mention that this technique requires a reliable online error control.

Well-Conditioned AWE

As already stated, PVL can only be applied to linear systems. This implies that lossy systems as defined in (2.3.12) should be linearized according to (2.3.14) in order to be reduced. Orthogonalizing the vectors $\mathbf{x}_i$ in the AWE algorithm (3.3.19) is not recommended as the moments would not match anymore. As already stated several methods have been proposed to improve the performance of this algorithm. Among them, the well-conditioned AWE (WCAWE) has been shown to be the most robust one [56].

The WCAWE iteration for the transfer function in (3.3.14) is given for MIMO systems as follows:

$$
\begin{aligned}
\widetilde{\mathbf{X}}_0 &= \hat{\mathbf{A}}_0^{-1}\hat{\mathbf{B}} \\
\widetilde{\mathbf{X}}_1 &= -\hat{\mathbf{A}}_0^{-1}\hat{\mathbf{A}}_1\mathbf{X}_0 \\
&\;\;\vdots \\
\widetilde{\mathbf{X}}_p &= \hat{\mathbf{A}}_0^{-1}(-\hat{\mathbf{A}}_1\mathbf{X}_{p-1} - \hat{\mathbf{A}}_2\mathbf{X}_{p-2}\mathbf{P}_{\mathbf{U}_2}^{-1}(n,2)\mathbf{E}_{p-2}),
\end{aligned}
\tag{3.4.7}
$$

where $\mathbf{E}_{p-2}$ is a $n \times 2$ matrix with its lower quadrant being the 2×2 identity matrix while all other entries are 0. The matrices $\widetilde{\mathbf{V}}_p = [\widetilde{\mathbf{X}}_0, \widetilde{\mathbf{X}}_1, \ldots, \widetilde{\mathbf{X}}_p]$ and $\mathbf{V}_p = [\mathbf{X}_0, \mathbf{X}_1, \ldots, \mathbf{X}_p]$ and related through the QR decomposition as illustrated in (3.3.20), where $\mathbf{U}$ is an upper triangular matrix. The adjustment terms, as far as $\mathbf{U}$ is nonsingular, are given as follows

$$
\mathbf{P}_{\mathbf{U}_i}(n,m) = \prod_{t=i}^{m} \mathbf{U}_{[t:n-m+t-1,t:n-m+t-1]}^{-1},
\tag{3.4.8}
$$

where $\mathbf{U}_{[\alpha:\beta,\alpha:\beta]}$ is a sub-matrix of $\mathbf{U}$ with columns and lines from α to β.

Partial Realization

Instead of the Taylor coefficients, the partial realization [57, 58] aims to approximate the Markov parameters. After a variable change from s to $z = 1/s$, we obtain the following expression for the transfer function:

$$
\mathbf{H}(z) = z\mathbf{C}(\mathbf{I} + z\mathbf{A})^{-1}\mathbf{B}.
\tag{3.4.9}
$$

Again by developing this expression in geometric series as in (3.3.7), we obtain

$$
\mathbf{H}(s) = \sum_{k=0}^{\infty} \underbrace{\mathbf{C}(-\mathbf{A})^k\mathbf{B}}_{\mathbf{M}_k} \frac{1}{s^k},
\tag{3.4.10}
$$

where $\mathbf{M}_k$ are the Markov parameters. The reduced system built by applying the following projection:

$$
\begin{aligned}
\mathbf{H}_p(s) &= \mathbf{C}\mathbf{V}_p(s\mathbf{W}_p^T\mathbf{V}_p + \mathbf{W}_p^T\mathbf{A}\mathbf{V}_p)^{-1}\mathbf{W}_p^T\mathbf{B} \tag{3.4.11a} \\
&= \mathbf{C}_p(s\mathbf{I} + \mathbf{T}_p)^{-1}\mathbf{B}_p, \tag{3.4.11b}
\end{aligned}
$$

where the matrices $\mathbf{V}_p$ and $\mathbf{W}_p$ result respectively from the Krylov subspaces $\mathcal{K}_p(\mathbf{A}, \mathbf{B})$, and $\mathcal{K}_p(\mathbf{A}^T, \mathbf{C})$ is a partial realization of the original system. In fact, it has been proven in [57] that the first $2m_p$ Markov parameters of $\mathbf{H}_p$ and $\mathbf{H}$ are identic.

This reduction process is a moment matching at $z = 0$ ($s \to \infty$). Thus, the partial realization is a particular case of the Padé approximation with an expansion point at infinity. The method, implemented in [10], is well suited for FIT systems as only matrix-vector multiplications are performed in order to build the Krylov space $\mathcal{K}_p(\mathbf{A}, \mathbf{B})$. Whereas at least one inversion of the matrix $\hat{\mathbf{A}}$ is needed in the Padé approximation. This is made possible by the diagonal form of the material matrices in FIT.

Applying it to FEM systems will not yield the same benefit, as the mass matrix (1.1.1) in the FEM state space representation is not of diagonal form. In fact, operating the same variable change as in (3.4.9) yields

$$\mathbf{H}(z) = z\mathbf{C}(\mathbf{I} + z\mathbf{E}^{-1}\mathbf{A})^{-1}\mathbf{E}^{-1}\mathbf{B}. \tag{3.4.12}$$

Thus, the computation of the related Krylov space $\mathcal{K}_p(\mathbf{E}^{-1}\mathbf{A}, \mathbf{E}^{-1}\mathbf{B})$ requires also to solve a system of the original size.

The main drawback of the partial realization is its slower convergence rate. It is obvious that interpolating at infinity is not as efficient as choosing the interpolation point in the frequency of range of interest. This results in Krylov spaces and thus reduced systems of higher dimension compared to the Padé approximation[7]. In order to tackle this point, the so-called *two step Lanczos* (TSL) was proposed in [10]. It consists of reducing the original system with less complexity through the partial realization and applying the Padé approximation to the then obtained systems in order to compute models of smaller order. In order to keep the storage requirements low, the Krylov spaces were built following the Bi-Lanczos algorithms.

While this method performs very well for systems of middle complexity (10^4 to 10^5 unknowns), it becomes less efficient for systems with more unknowns and resonances for two reasons:

- As the system is approximated from infinity, more vectors are needed to catch the resonances lying in the considered frequency range which are essential for accurate results.

---

[7]The dimension of the reduced systems resulting from a partial realization are of factor higher than 100 compared to those obtained with the Padé approximation according to the results obtained in [10]

- As stated in [47], the orthogonality of the Krylov spaces built with the Bi-Lanczos algorithm is lost due to numerical errors in spaces with hundreds of vectors. Thus, mitigating the storage requirements by keeping only a few vectors as in [10] would lead to numerical errors and thus lower convergence rates or even stagnation.

This strengthens the fact that the Padé approximation is the most suitable method for very large scale problems even in FIT systems. Furthermore, keeping all the Krylov space vectors enables the computation of field vectors.

3.5 Passive Reduction

The order reduction through Padé approximation requires the asymmetric projection of the original system onto the matrices $\mathbf{V}_p$ and $\mathbf{W}_p$ (3.4.11). The reduced systems are then no more guaranteed positive real matrices and thus, cannot assure the conservation of stability and passivity according to Section 2.5. While these properties are not of relevance in frequency domain computations, they are indispensable for time domain calculations. Thus, as already stated, only passive reduced models can be considered in the scope of time-domain field-circuit simulations.

Several methods for passive model order reduction have been presented in the past [59–62]. Among them, the so-called *passive reduced interconnect macromodeling algorithm* (PRIMA) [59] has emerged as one of the mostly used for the Padé approximation. It consists of projecting the original system onto the orthonormal matrix $\mathbf{V}_p$ resulting from the Krylov space $\mathcal{K}_p(\hat{\mathbf{A}}, \hat{\mathbf{B}})$:

$$\mathbf{H}_p(s) = \mathbf{B}^T \mathbf{V}_p (s \mathbf{V}_p^T \mathbf{V}_p + \mathbf{V}_p^T \mathbf{A} \mathbf{V}_p)^{-1} \mathbf{V}_p^T \mathbf{B} \tag{3.5.1a}$$

$$= \mathbf{B}_p^T (s \mathbf{I} + \mathbf{A}_p)^{-1} \mathbf{B}_p, \tag{3.5.1b}$$

where we use the fact that in our case, $\mathbf{C} = \mathbf{B}^T$. The reduced matrix $\mathbf{A}_p$ is neither tridiagonal nor of Hessenberg form as the projection matrix is computed with the matrix $\hat{\mathbf{A}}$. Assuming a real s_0, this symmetric projection, also called congruence transformation, guarantees obviously the preservation of passivity. In fact, the reduced system (3.5.1) is equivalent to the general form (2.5.5) and thus passive as $(s\mathbf{I} + \mathbf{A}_p)$ is positive real.

Only the first m_p moments of the PRIMA reduced and original systems match together:

$$\hat{\mathbf{M}}_k = \mathbf{B}_p^T \check{\mathbf{A}}_p^{k+1} \mathbf{B}_p \qquad \text{with } \check{\mathbf{A}}_p = (s_0 \mathbf{I} + \mathbf{A}_p)^{-1} \tag{3.5.2a}$$

$$\hat{\mathbf{M}}_k = \mathbf{M}_k \qquad \text{for } 0 \leq k < m_p - 1, \tag{3.5.2b}$$

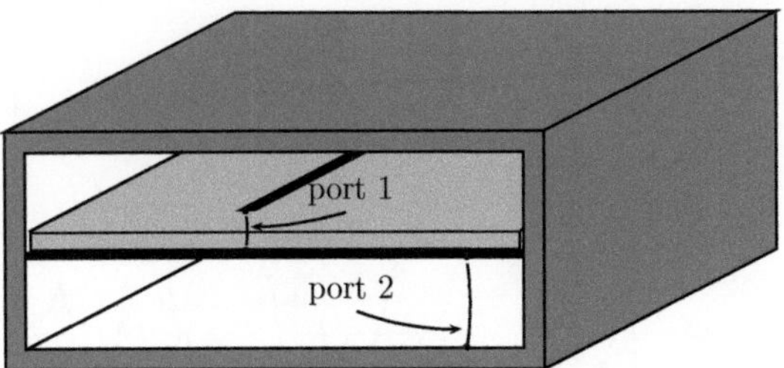

Figure 3.2: Model for housing concept analysis

as proved in [59, 60]. Thus, the conservation of passivity is guaranteed at cost of accuracy. However, the number of matching moments gives only an estimation of the accuracy level so that in some cases doubling the number of moments may not improve the accuracy in the same order.

In order to illustrate the efficiency of the implemented PRIMA method, the test structure represented in Fig. 3.2 has been analyzed. It consists of a printed circuit board (PCB) modeled as a simple two-layer board within its housing. This model aims to study the impact of different housing concepts on the EMC behavior of the PCB. For this purpose, two ports were considered:

- The first port which is placed between the trace and the PCB ground represents the excitation.

- The second port is required to compute the coupling between PCB ground and metal housing.

The considered frequency interval is from 1 MHz to 4 GHz and the discretization yields 154,000 unknowns in the curl-curl formulation. This example will be also considered for following illustrations in this section and in Chapter 4.

Fig. 3.3 shows the comparison of the relative error $E_{i,j}$ of Z-parameter entries, defined as follows

$$E_{[i,j]}(s) = \frac{\|Z_{p[i,j]}(s) - Z_{[i,j]}(s)\|}{\|Z_{[i,j]}(s)\|}, \qquad (3.5.3)$$

at different dimensions p of the Krylov space (20, 40 and 60) with the expansion frequency at 1 GHz for $Z_{[1,2]}$, the impedance between the ports 1 and 2. The local

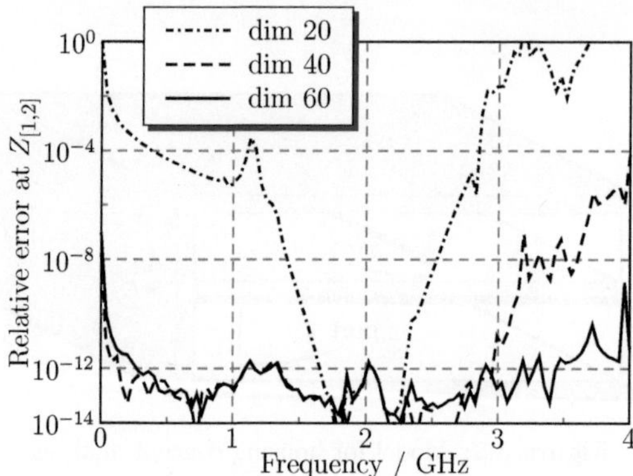

Figure 3.3: Relative error at $Z_{[1,2]}$ for the PCB-housing model of the PRIMA systems at dimensions 20, 40 and 60 for PCB-housing model

approximation properties of moment matching methods in general and Padé approximation with imaginary interpolation point particularly as mentioned in Section 3.4.2 can be observed. In fact, the convergence bandwidth of the reduced system is first close to the expansion point and grows gradually with increasing Krylov spaces. The relative error drops to 10^{-8} in the whole frequency range already with a reduced dimension of 60.

The results obtained with our MOR code at dimension 56 are compared with those of Microwave Studio (MWS®) from CST [63] from which we got the material matrices, $\mathbf{M}_\varepsilon$ and $\mathbf{M}_\mu$, in Fig. 3.4. It can be clearly seen that the results obtained from MOR agree very well with those from MWS®. The high speedup in time of 15 is due to the fact that only one inversion is required for MOR and 36 for the frequency sweep of MWS® as the structure is very resonant. This speedup can be even improved to 30 by considering only an accuracy of 10^{-2} which is achieved with 30 Krylov vectors. The memory size, the number of inversions, and the solver time are summarized in Table 3.5.

3.6 Generation of Macromodels

As already stated in Section 2.1, Maxwell's equations give a general description of all electromagnetic phenomena even in domains with complex and inhomogeneous

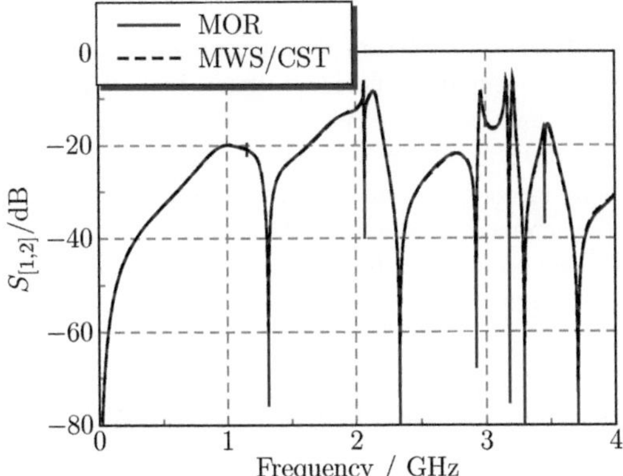

Figure 3.4: Comparison of $S_{[1,2]}$ for PCB-housing model between MOR and MWS® (FD evaluation)

	MWS®	MOR
method	FD (frequency sweep)	passive Padé
# unknowns	154, 000	154, 000
memory size	520 MB	312 MB
# inversions	36	1
accuracy	10^{-4}	10^{-4}
solver time	13 min.	**51s.**

Table 3.5.1: Solver comparison between MOR and MWS® for PCB-housing model

material distribution. On the other hand, nodal analysis is a strong abstraction of Maxwell's equations based on the currents, voltages and lumped elements. Voltages and currents are related with each other through constitutive equations like the Ohm's law.

Nodal analysis assumes that the elements are smaller than the considered wavelengths and connected with perfect conductors with vanishing signal delay. Despite this coarse approximation, nodal analysis can handle a wide variety of circuits with sufficient accuracy. Furthermore, it enables the computation of complex and nonlinear circuits which would go beyond the limits of field simulation.

As effects like cross-talking, electromagnetic radiation, delay and dispersion which cannot be neglected due to increasing operating frequencies require the consideration of Maxwell's equations, coupling of field and circuit simulation is indispensable. A direct coupling in time domain may be inefficient as the ratio between the time stepping for the field simulation[8] and the circuit simulation time[9] would lead to high computation times especially for long field simulations. Therefore, the electromagnetic effects are described as macromodels which can be plugged into a circuit simulation, i.e SPICE [64]. This may occur either by direct stamping of the elements of the reduced models in SPICE circuits or through a Y-parameter description of the macromodels.

Macromodel extraction from measurement or simulation data has been a topic of intensive research in the past. Several robust methods from the system identification theory have been proposed [65–68]. The extracted models are however not guaranteed passive e.g. due to coarse frequency sampling. While finer sampling would tremendously increase the simulation time, passivity enforcement [69, 70] may require some compromises regarding accuracy. MOR can be considered as an efficient option by providing accurate and guaranteed passive models.

3.6.1 Modified Nodal Analysis

The modified nodal analysis (MNA) [71] is an extension of the classical nodal analysis [35] in the sense that it allows to consider independent current and voltage sources and mutual inductances in a network. Considering a network with predefined nodes and branches and applying the Kirchoff's current law (KCL) and branch constitutive equations, we obtain the following system

$$\begin{pmatrix} \mathbf{G}+s\mathbf{C} & \mathbf{W}^T \\ -\mathbf{W} & s\mathbf{L} \end{pmatrix} \begin{pmatrix} \mathbf{v} \\ \mathbf{j} \end{pmatrix} = \begin{pmatrix} \mathbf{B} \\ \mathbf{0} \end{pmatrix} \mathbf{i}, \qquad (3.6.1)$$

where $\mathbf{v}$ and $\mathbf{j}$ are the vectors containing the nodal voltages and the currents at the inductors, respectively. The matrices $\mathbf{G}$, $\mathbf{C}$ and $\mathbf{L}$ contain the stamps for resistors, capacitors and inductors respectively. The matrix $\mathbf{W}$ consists of 1, -1 and 0 and $\mathbf{B}$ is the coupling matrix to the port current $\mathbf{i}$. Through the transformation

$$\mathbf{j} = \frac{1}{s}\mathbf{L}^{-1}\mathbf{W}\mathbf{v}, \qquad (3.6.2)$$

---

[8]Some ns in structures with small details

[9]Typically some ms due to slow transients

and elimination of $\mathbf{j}$ the state space representation can be reduced to the expression

$$\left(\frac{1}{s}\underbrace{\mathbf{W}^T\mathbf{L}^{-1}\mathbf{W}}_{\mathbf{Y}_L}+\mathbf{G}+s\mathbf{C}\right)\mathbf{v}=\mathbf{Bi} \qquad (3.6.3a)$$

$$\mathbf{u}=\mathbf{B}^T\mathbf{v}, \qquad (3.6.3b)$$

with $\mathbf{u}$ being the voltages at the ports. The matrices $\mathbf{G}$, $\mathbf{C}$ and $\mathbf{Y}_L$ are symmetric semi positive definite and diagonal dominant. Their diagonal entries are positive whereas the other are negative.

3.6.2 Direct Realization

By considering the reduced system from the general state space representation in (3.3.14)

$$(s\mathbf{I}+\mathbf{A}_{p1}+1/s\mathbf{A}_{p0})\mathbf{X}_p=\mathbf{B}_p\mathbf{i} \qquad (3.6.4a)$$

$$\mathbf{u}=\mathbf{C}\mathbf{X}_p, \qquad (3.6.4b)$$

and comparing with the system resulting from MNA (3.6.3), the analogy is obvious. In fact, $\mathbf{I}$, $\mathbf{A}_{p1}$, and $\mathbf{A}_{p0}$ are equivalent to $\mathbf{C}$, $\mathbf{G}$, and $\mathbf{Y}_L$, respectively. For the realization in SPICE models, we will distinguish between systems with an without losses.

Lossless case

In this case, the state space is given as

$$(s\mathbf{I}+1/s\mathbf{A}_p)\mathbf{X}_p=\mathbf{B}_p\mathbf{i} \qquad (3.6.5a)$$

$$\mathbf{u}=\mathbf{B}_p^T\mathbf{X}_p. \qquad (3.6.5b)$$

The matrices $\mathbf{A}_p$ resulting from the PRIMA process are dense and would thus lead to a SPICE circuit with the maximum of elements $(p(2m+p))$ which is of order $\mathcal{O}(p^2)$. The diagonalization is one of the most popular methods to reduce this amount of circuit elements. After the eigendecomposition $\mathbf{A}_p=\mathbf{V}\mathbf{\Lambda}\mathbf{V}^T$, (3.6.6b) can be derived to

$$(s\mathbf{I}+1/s\mathbf{\Lambda})\mathbf{X}_r=\underbrace{\overbrace{\mathbf{V}^T\mathbf{B}_p}^{\mathbf{i}_r}\mathbf{i}}_{\mathbf{B}_r} \qquad (3.6.6a)$$

$$\mathbf{u}=\mathbf{B}_p^T\mathbf{V}\mathbf{X}_p. \qquad (3.6.6b)$$

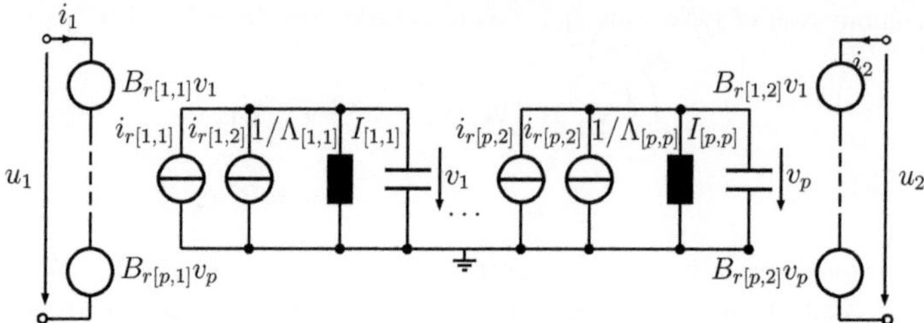

Figure 3.5: SPICE circuit resulting from the diagonalized reduced model

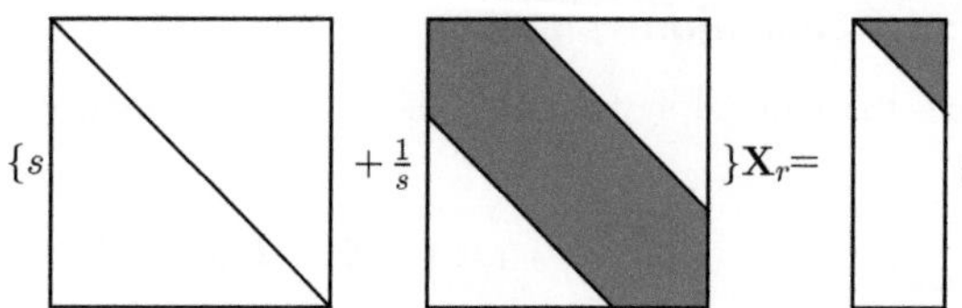

Figure 3.6: System after partial realization as second reduction step. The capacitor matrix is diagonal, the inductor matrix has m secondary diagonals, and the right hand side has m elements where m is the number of ports

The direct realization of (3.6.6) leads to a circuit with p nodes consisting of an inductor and capacitor in parallel, and the controlled currents $\mathbf{i}_r$, and the branches corresponding to the port voltages as illustrated in Figure 3.5. The number of elements, $p(2m + 2)$, is linear in p.

In nodal analysis, the dimension of the matrices is defined by the number of voltage sources in the circuit. Thus, the systems in (3.6.6) are enlarged by the number of sources at the port branches. This leads to high computation time especially while analyzing systems with high dimension and/or number of ports. In order to remedy to it, partial realization is run as second reduction step. By doing so, the number of voltages at each port branch is reduced to one and the number circuit elements is $p(2m + 2) - m^2 + 5m + 2$ (Fig. 3.6). In addition, the partial realization reduces the model further to yield macromodels of even smaller size.

The elements on the off-diagonals in Fig. 3.6 are modeled as inductors between the corresponding nodes as illustrated in Fig. 3.7. Note that the values of the inductors may be negative as the reduced matrices do not necessary have the same structure as MNA matrices. However, this would not affect the accuracy of the results as macromodels are just a mathematical description of the transfer function.

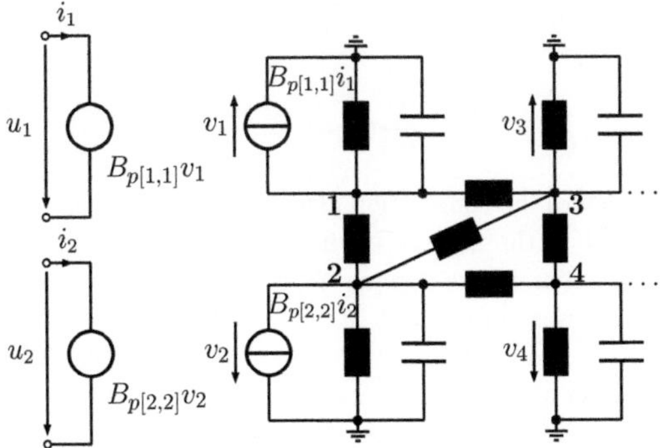

Figure 3.7: SPICE circuit resulting from partial realization as second reduction step

Lossy case

In the scope of this section we focussed on losses resulting from materials with finite conductivity or resistors as PML yield nonsymmetric matrices which are in contradiction with the properties of MNA matrices. The reduced systems are then of the form (3.6.4). Besides the inductor and capcitor matrices already discussed above, we can recognize the resistor component which stands for losses. Again, the model extraction can be performed as for cases without losses, except that resistors shall be considered additionally.

3.6.3 Y-Parameter Description

The Y-parameters of the reduced system can be represented, as already introduced for the more general transfer function (3.3.10), in rational polynomials

$$Y_{[i,j]}(s) = \frac{b_p s^p + \cdots + b_1 s + b_0}{a_{p-1} s^{p-1} + \cdots + a_1 s + 1}, \tag{3.6.7}$$

where p is the order of reduction. This expression can be directly used in a circuit simulation in frequency domain whereas a convolution is indispensable for transient simulation. It consists of transforming the Laplace relationship

$$\mathbf{I} = \mathbf{Y}(s)\mathbf{U} \tag{3.6.8}$$

between currents $\mathbf{Y}$ and node voltages $\mathbf{U}$ into the following integral expression

$$i_j(t) = \sum_{k=1}^{m} \int_0^t y_{j,k}(t-\tau)v_k(\tau)d\tau. \tag{3.6.9}$$

The main drawback of the convolution algorithm resides in its quadratic complexity, $\mathcal{O}(T^2)$ with respect to the number of time points during simulation, T. Therefore, recursive convolution and time domain Y parameters macromodels which guarantee a linear complexity were introduced [72, 73]. Even though this method is more efficient as the direct realization we could not test it as SPICE3f4 [74] in which it is implemented was not available.

4 Model Order Reduction for EMC Purposes

The main focus of this work is the application of MOR on EMC problems, which is the topic of this chapter. The demand for high qualitative telecommunication and reliable safety standards in cars have led to more restrictive EMC norms. EMC simulation has become indispensable in early development phases due to tremendous improvements both on the hardware and the software side. In order to improve the practicability of these complex computations, some efficient methods in combination with MOR will be proposed in this chapter. The outline follows the framework presented in the introduction:

1. *EMC computation in which we will present the error control method implemented in this work as well as a method to reduce the dimension of the reduced systems in presence of a lot ports. Furthermore, we propose a method for parallelization and discuss the efficiency of WCAWE with PML boundaries.*

2. *EMC analysis enables a derivation of equivalent circuit models in order to determine the causes of deviations from the EMC norm.*

3. *EMC optimization in which a stand-alone optimization workflow which combines a genetic algorithm and MOR is proposed.*

4.1 EMC Computation

In order to predict the EMC properties of an electronic device, several methodologies may be considered. Either electromagnetic interactions of different parts of the device should be investigated or signals propagating on electrical circuits are of interest. In most cases, the combination of both is recommended. In fact, the frequency of interest limits the accuracy of circuit simulation, whereas the consideration of some measurement setups and/or nonlinear elements in the implemented electrical circuits enforce it.

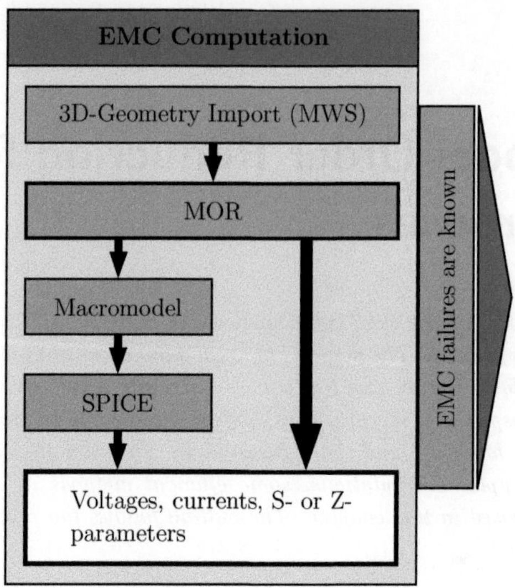

Figure 4.1: Overview of the block EMC computation.

EMC computation may then consist of calculating the transfer function in terms of Z- or S-parameters or computing relevant current and/or voltages in a circuit simulation with a macromodel resulting from the field simulation. It has already been shown in the last chapter that MOR is a robust method to achieve this goal. For this purpose, a C++ code has been implemented. The discretization is performed with MWS$^®$ which provides the material matrices ($\mathbf{M}_\varepsilon$, $\mathbf{M}_\mu$, and $\mathbf{M}_\kappa$) and mesh information necessary for the setup of the FIT system. The output of this code consists of the S-, Z-parameters, and a SPICE macromodel which can be used in a circuit simulation. This block allows to identify whether some EMC failures shall be expected for the structure under test by analyzing predicted voltages, currents or the transfer function (Figure 4.1).

The efficiency of MOR strongly depends on a reliable error control[1]. In fact, the reduced systems should not be too small as the accuracy is no more guaranteed. On the other hand, the size should not exceed its optimum as the efficiency would be degraded. Two methods will be presented in this section in order to address this

---

[1]An error estimation supposes an error bound which is not defined for the Padé approximation presented in this work.

issue. It has also been mentioned that the size of the reduced models increases with the number of ports. Thus, for systems with massive number of ports (> 30 ports)[2], a method which consists of the combination of modal truncation and Padé approximation is proposed. Furthermore, increasing model order combined with improvements for cluster solutions require the consideration of parallelization methods.

4.1.1 Error Control

The main inconveniency of moment matching methods is the lack of a strict online error control. Several methods have been introduced to address this issue [52, 55, 75, 76]. We will mainly focus on convergence monitoring and a residual based method.

Residual-based Method

The residual is an essential stop criterion parameter in iterative linear solvers [47]. Considering the following system of equations

$$\mathbf{A}\mathbf{x} = \mathbf{b},\tag{4.1.1}$$

the residual at iteration step i with the approximated solution $\mathbf{x}_i$ is defined as

$$\mathbf{r} = \mathbf{A}\mathbf{x}_i - \mathbf{b}.\tag{4.1.2}$$

The error is related to the residuum as follows

$$\mathbf{e} = \mathbf{x}_i - \mathbf{x}\tag{4.1.3a}$$
$$= \mathbf{A}^{-1}\mathbf{r}.\tag{4.1.3b}$$

Monitoring the residual over the iteration steps thus gives a trend of the convergence rate of the solver. In fact, small errors imply small residuals. This knowledge has been introduced in [52] to derive a stop criterion for Lanczos and Arnoldi related MOR methods. The residual of reduced models following Arnoldi or Lanczos algorithms is given as follows

$$\mathbf{R}(s) = \underbrace{(\mathbf{A} + s_0\mathbf{I})}_{\mathbf{A}_{s_0}}(\mathbf{I} + \hat{s}\hat{\mathbf{A}})\widetilde{\mathbf{X}} - \mathbf{B}\tag{4.1.4a}$$
$$= \mathbf{A}_{s_0}\left((\mathbf{V}_p + \hat{s}\hat{\mathbf{A}}\mathbf{V}_p)\mathbf{X}_p - \hat{\mathbf{B}}\right), \text{ with } \hat{\mathbf{B}} = (\mathbf{A} + s_0\mathbf{I})^{-1}\mathbf{B}\tag{4.1.4b}$$
$$= \mathbf{A}_{s_0}\mathbf{V}_p\underbrace{\left((\mathbf{I} + \hat{s}\mathbf{T}_p)\mathbf{X}_p - \hat{\mathbf{B}}_p\right)}_{=\mathbf{0}} + \hat{s}\mathbf{A}_{s_0}\mathbf{W}_p\mathbf{X}_p.\tag{4.1.4c}$$

---

[2]In ICs (integrated circuits), the number of pins easily exceeds this value.

From (4.1.4b) to (4.1.4c) we used the Arnoldi's property illustrated in Fig. 3.1. The vector $\tilde{\mathbf{X}} = \mathbf{V}_p \mathbf{X}_p$ represents the approximated state space variables with

$$\mathbf{X}_p = (\mathbf{I}_p + \hat{s}\mathbf{T}_p)^{-1}\hat{\mathbf{B}}_p \tag{4.1.5a}$$

$$\approx (\mathbf{A}_p + s\mathbf{I}_p)^{-1}\mathbf{B}_p, \tag{4.1.5b}$$

being the state space vector in the reduced model. This explains also why the expression in 4.1.4c is 0. We assume that $\mathbf{X}_p$ should be approximatively the same no matter if the one-side projection has been performed on the original system (3.3.2) or the shifted in (3.3.4), so the expression for the residual in (4.1.4c) can be used for models reduced with PRIMA.

As already pointed out in [52], the residual does not provide an estimation of the error which requires the inversion of the original system $(\mathbf{A} + s\mathbf{I})$. Therefore, we introduce the following expression for the approximated error as defined in (4.1.3b)

$$\widetilde{\mathbf{E}}(s) = \hat{s}\mathbf{B}_p^T(\mathbf{A}_p + s\mathbf{I}_p)^{-1}\mathbf{V}_p^T\mathbf{A}_{s_0}\mathbf{W}_p\mathbf{X}_p, \tag{4.1.6}$$

where the reduced system is inverted instead of the original one[3]. By doing so, the computation complexity of the approximated error can be kept very low. On the other hand, this estimation is not an error bound but yields a good approximation as illustrated in Figures 4.2, 4.3, and 4.4.

Again, we considered the example introduced in Section 3.5. The relative approximated error is compared with the exact one at different dimensions (20, 40, and 60). It can be seen that the residual-based approximated error is mostly above the exact one. It should be stated that relative errors below 10^{-12} are no more relevant as they are beyond numerical accuracy. Even though there is no mathematical proof for this efficiency, it can be explained as follows:

- The reduced model is an approximation of the original one, thus the higher the dimension, the better the approximation of the error following (4.1.6).

- The residual is so high in frequency intervals where the reduced model has not yet converged, so that it dominates the approximated error there.

This error control was indeed differently derived in [75] for PRIMA systems but the main issue of our derivation is the obvious relation between residual and error, and thus the consistence of the method. On the other side, we were able to apply it on Maxwell's equation systems resulting from FIT discretization.

---

[3] $(p \times p) \ll (N \times N)$

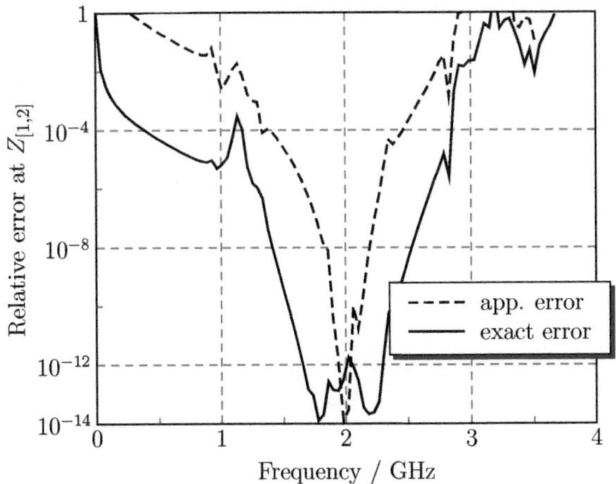

Figure 4.2: Approximated error vs. exact error over the frequency for the PCB-housing model at a Krylov dimension of 20

Convergence Monitoring

One of the drawbacks of the error control introduced above is the fact that it can only be applied on linear systems. Though *curl-curl* systems with losses reduced with WCAWE (Section 3.4.2) cannot be addressed. For this purpose, we need a more general ansatz.

The convergence monitoring consists of approximating the real error $\mathbf{E}(s)$ through

$$\widetilde{\mathbf{E}}(s) = \mathbf{H}_p(s) - \mathbf{H}_q(s), \tag{4.1.7}$$

where $\mathbf{H}_p(s)$ and $\mathbf{H}_q(s)$ are systems obtained from two different reduction schemes. The first variant, introduced in [52], consists of choosing two complementary reduced systems of the same size $\mathbf{H}_p(s)$ and $\mathbf{H}_{p\perp}(s)$. The two different approximations assume that they agree at frequencies where $\mathbf{E}(s)$ is small. The complementarity is achieved by interpolating at different frequencies. Although the results are satisfying, the increased complexity by computing a LU decomposition at two frequencies instead of one is not worth.

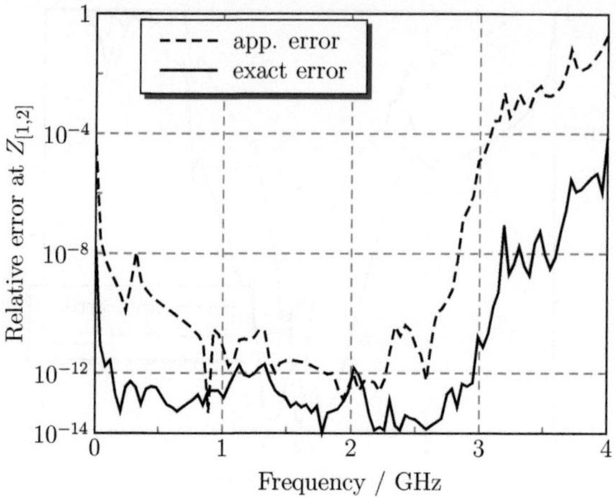

Figure 4.3: Approximated error vs. exact error over the frequency for the PCB-housing model at a Krylov dimension of 40

Another more practicable method has been therefore presented in [76]. For this purpose, we consider an exact value x which shall be approximated by the sequence x_n. The error $e_n = |x_n - x|$ at any iteration step n can be given in the following expressions

$$|x_n - x| = |x_n - x_{n-1} + x_{n-1} - x| \geq |x_n - x_{n-1}| - |x_{n-1} - x|, \qquad (4.1.8a)$$
$$= |x_n - x_{n-1} + x_{n-1} - x| \leq |x_n - x_{n-1}| + |x_{n-1} - x|. \qquad (4.1.8b)$$

With $\tilde{e}_n = |x_n - x_{n-1}|$, the inequalities above can be combined in

$$|e_{n-1} - e_n| \leq \tilde{e}_n \leq e_{n-1} + e_n. \qquad (4.1.9)$$

It has been observed in practice that $\tilde{e}_n$ can be well approximated by e_{n-1} and thus, the approximated error is given as

$$\widetilde{\mathbf{E}}(s) = \mathbf{H}_p(s) - \mathbf{H}_{p-1}(s). \qquad (4.1.10)$$

Again this expression assumes that if the difference by increasing the dimension of the Krylov space gets smaller then the reduced system has converged and the error $\mathbf{E}(s)$ is also small.

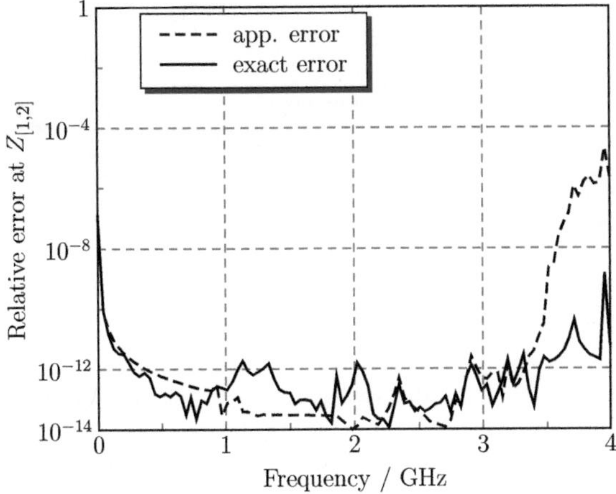

Figure 4.4: Approximated error vs. exact error over the frequency for the PCB-housing model at a Krylov dimension of 60

All these methods suffer from the fact that they are not consistent. In fact, if the two systems in (4.1.7) have not yet converged, then the error may be wrong. The same holds for (4.1.10).

Stop Criterion

In order to provide a reliable stop criterion, the error should be approximated at several frequencies in the considered range. In this work, it has been observed that the error is well monitored with a value of 100 frequency samples equally spread over the range. Assuming the estimation is performed in one of the two methods presented above, the figure of merit for convergence, e, is set as

$$e = \frac{\sum_{i=1}^{100} \max\{\widetilde{\mathbf{E}}(s_i)\}}{100}.$$

(4.1.11)

In this way, the figure of merit is not dominated by error peaks which usually occur close to resonances but consider the overall convergence. In fact, an accuracy of 1% in the detection of resonances is good enough in EMC context. At the contrary to

Figure 4.5: Wire model with two ports above a conducting table used to investigate the efficiency of MOR in presence of PML boundaries.

antenna analysis, the cause of the resonance and not its location with high definition is of interest.

In practice, the error is not approximated after each iteration, the user sets an initial dimension for the reduced model and an incremental value after which e is compared to a threshold value ε_{th}. If $e < \varepsilon_{th}$, the computation is stopped and the reduced model is retrieved. A threshold value of $\varepsilon_{th} = 10^{-4}$ has been found to be a good compromise between accuracy and efficiency. It should be stated that even an accuracy of 10^{-2} is acceptable for EMC purposes. This threshold may further improve the efficiency of MOR compared to common frequency sweeps.

4.1.2 MOR with PML

In order to investigate the performance of the implemented MOR method in presence of PML boundaries, we considered the following setup illustrated in Figure 4.5:

- a 1 m long wire,

- a perfect conducting table 5 cm under the table,

- two ports at the ends of the wire.

Apart from the lower side which is terminated with PEC, the boundaries of the computational domain have been modeled as 6 PML layers. The minimal reflection factor has been set to 10^{-4} and the exponent of the geometric function (2.2.24) is 3. Furthermore, the discretization leads to $3 \cdot 10^5$ DOFs.

Figure 4.6 shows the comparison of the transmission ($S_{[1,2]}$) between MWS[®][4] and MOR for a Krylov dimension of 1000. It can be clearly seen that the reduced system

---

[4]The boundaries properties have been set as for the MOR computation.

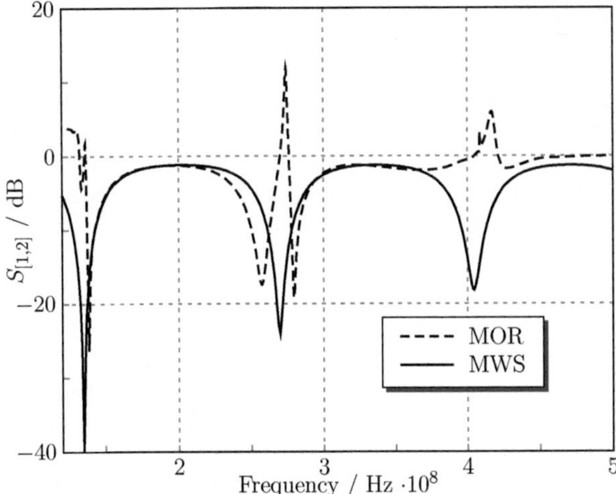

Figure 4.6: $S_{[1,2]}$ of the wire model with PML boundaries obtained from MOR with a Krylov dimension of 1000 compared to the results from MWS®.

is still far from convergence. As a comparison, a model of the same order without PML can be reduced to a system of dimension 20. This is due to the fact that PML layers worsens the condition of the system matrices. It has been showed in [77] that solving FIT matrices iteratively considering PML requires 600 times more iterations than without PML.

The Padé approximation can be considered as an iterative method to resolve the main eigenvalues of the system in the neighborhood of the expansion point. In this way, the bad condition of the matrix has the same effect on the convergence rate as in [77]. The condition number can surely be improved by not considering the PML elements of higher order but there is still a factor 10 in the number of iterations compared to systems without PML as investigated in [77]. Therefore, it is not worthwhile to apply MOR for systems with PML. It is recommendable to consider Mur boundaries which of course are less accurate than PML but do not affect the condition of the matrices.

4.1.3 Modal Truncation + Padé

In some applications where ICs are involved, the number of ports related to the pins under test may get beyond 30. This represents a big challenge for moment matching methods. Especially for resonant structures where a high number of moments should be matched for a good accuracy: the size of the reduced model gets so huge that the term order reduction is no more appropriate. Not only the memory requirements to store the Krylov vectors grows tremendously but also the computation with the macromodels gets inefficient.

Several methods have been proposed to address this issue:

- The port correlation can be used in order to reduce the dimension of the Krylov space. For this purpose a singular value decomposition of the transfer function $\mathbf{H}(s)$ is performed at a given frequency and used to represent the set of ports as a small-rank approximation (**SVD-MOR** [78]). The reduction is applied on this small-rank approximation and then projected to the original system by using the singular vectors. This approximation can be also frequency-dependent [79].

- The **decentralized MOR** [80] aims to split the $m \times m$ MIMO systems into $1 \times m_p$ SIMO systems where m_p is reduced by eliminating ports with less interaction. The port interactions are quantified by computing the relative gain array (RGA) of the transfer function. If $m_p \ll m$, and the number of matched moments is the same, then the computation of Z-parameters would be less time and memory consuming.

In the cases considered in this work, the transfer function is of full-rank which degrades the efficiency of SVD-MOR. Furthermore, the number of matched moments for a given accuracy of the reduced models depends also on the number of ports. In fact, increasing the number of ports would require less moments for convergence. This can be illustrated by considering the analogy to iterative eigenvalue solvers. The more search vectors are used, the less iterations are needed for satisfying convergence. This discards decentralized MOR which assumes a constant number of moments for high efficiency.

Therefore, the combination of modal truncation and Padé approximation was considered to reduce MIMO systems with a high number of ports, first presented in [81]. Any transfer function can be represented through its poles and zeros. The modal analysis leads to the truncated system introduced in (3.2.14b)

$$(\mathbf{V}_1^T \mathbf{A} \mathbf{V}_1 + s\mathbf{I})\mathbf{x}_1 = \mathbf{V}_1^T \mathbf{B} \mathbf{i} \tag{4.1.12a}$$

$$\mathbf{u}_1 = \mathbf{B}^T \mathbf{V}_1 \mathbf{x}_1, \tag{4.1.12b}$$

where $\mathbf{V}_1 \in \mathbb{R}^{N \times p}$ represents the eigenvectors corresponding to the dominant eigenvalues of the system under test.

As already stated in Section 3.2.3, this truncation is inaccurate and should be enhanced with a correction term. In fact, the modal analysis approximates solely the poles of the systems. The correction is, unlike presented in Section 3.2.3, computed with a Padé approximation. The matrix $\mathbf{V}_2$ which spans the Krylov space $\subset \mathcal{K}_r(\hat{\mathbf{A}}, \hat{\mathbf{A}}\mathbf{B})$ is then appended to $\mathbf{V}_1$ to form the orthonormal projection matrix $\mathbf{V}_q$

$$\mathbf{V}_q = [\mathbf{V}_1, \mathbf{V}_2]. \tag{4.1.13}$$

Afterwards, the reduced system is obtained through projection, while this is achieved through a two side projection in [81], we chose again the congruence transformation for guaranteed passive models.

By doing so, the main drawbacks of both methods can be compensated while enhancing their respective advantages. The poles of the system which are independent of the number of ports can be computed with the modal analysis. While the zeros which are strongly related to the ports are efficiently obtained through Padé approximation. If all poles and their corresponding eigenvectors in the considered frequency range and its neighborhood are known, then less moments are necessary to reduce the system with the Padé approximation.

Iterative solvers are the most efficient way to perform modal analysis with methods such as Jacobi-Davidson or Arnoldi [43]. In order to improve the convergence of those iterative methods, it is indispensable to regularize the FIT system as discussed in Section 2.4.3.

This method has been tested on a simple model of 25 conductors with square-shaped cross-sections. Even though the method proposed in this section requires $\frac{1}{2}$ to $\frac{1}{3}$ less vectors as the classical passive Padé approximation, it is still not enough to justify the time consuming eigenvalue computation. However, if the memory capacity is an issue, then the combination modal analysis and Padé approximation should be considered as worthwhile alternative.

4.1.4 Parallelization

Introduction

The simulation of complex structures ($> 10^7$ unknowns) needs tremendously high computation times and may even not be computed on standard personal computers (PCs) because of high memory requirements. *High performance computing* has been introduced for this purpose. In fact, hardware and software resources provided for parallelization enable faster computation of huge problems.

Parallel computers can be subdivided into the following categories [82]:

- **Multi-processor** machines have several processors which can access the same RAM (random access memory) through a memory bus. These so-called *shared memory machines* (SMM) enable an easier implementation as the memory management is mainly hardware-related and provide a reduced latent time for memory access. However, the number of processors in such systems is limited due to cost and complexity factors [83]. The main task for the programmer consists of guaranteeing a data consistence through synchronization of operations as they are performed on the same RAM. In this context, OpenMP has been established as the most reliable interface for implementation purposes [84].

- **Multi-computer** systems consist of several computers with their processors having their own RAM. Also called *distributed memory machines* (DMM), they are connected through a network bus which each other. Computations with these systems require data transfer within the involved computers and depending on the network performance, this may be very time consuming. On the other side, as the memory management is not supported by the hardware, the software implementation in this environment gets more complex. Nevertheless, this variant offers the best trade-off between hardware cost and computation performance due to continuous price fall of computer and network components coupled with their performance improvement. It is therefore the most widely used system for parallel computing [85]. In order to enable its implementation, MPI (message passing interface) has emerged as the most used interface [86].

Nowadays, hybrid models have grown to standards for parallel computing purposes. They are a combination of the two above presented models and consist of several multi-processor machines. In order to guarantee a reliable implementation in those systems, a combination of the interfaces OpenMP and MPI is indispensable.

The most relevant figures of merit of any algorithm are computation time T and memory requirement in relation to the problem complexity N and the number of processors N_π in the context of parallel computing. Besides enabling the computation of complex problems which cannot be performed on one machine, the parallelization aims to reduce the computation time, also known as scalability. Therefore, a figure of merit for parallel algorithms is their speedup which is defined as follows [87]

$$S(N_\pi, N) = \frac{T(1, N)}{T(N_\pi, N)}. \tag{4.1.14}$$

It represents the reduction factor of computation time between sequential $(T(1, N))$ and parallel implementation $(T(N_\pi, N))$. Another figure of merit for parallel algorithms is the efficiency

$$\epsilon(N_\pi, N) = \frac{T(1, N)}{N_\pi T(N_\pi, N)} = \frac{S(N_\pi, N)}{N_\pi}, \tag{4.1.15}$$

which describes the ratio between arithmetic operations and other tasks like data management and network communication.

Ideally, the optimal value for the speedup is

$$S_{opt} = N_\pi, \tag{4.1.16}$$

and therefore for the efficiency

$$\epsilon_{opt} = 1. \tag{4.1.17}$$

However, some discrepancies can be observed by analyzing explicitly the behavior of speedup and efficiency dependent on the number of processors under real conditions as shown in Figure 4.7. In fact, the computation time can be decomposed in a parallelizable $T_p(N_\pi, N)$ and a sequential part $T_s(N)$ which is executed by some or all in the computation involved processors

$$T(N_\pi, N) = T_s(N) + T_p(N_\pi, N). \tag{4.1.18}$$

Thus, the speedup can not exceed the value of

$$S_\infty(N) = \lim_{N_\pi \to \infty} S(N_\pi, N) = 1 + \frac{T_p(1, N)}{T_s(N)} \tag{4.1.19}$$

following the Amdahl's law [88]. In this theory, the speedup cannot increase linearly with the number of processors but approaches asymptotically the value in (4.1.19). The goal should thus be to parallelize any algorithm step for a better efficiency.

Another factor which affects the efficiency of a parallel algorithm is the communication between the processors. The more data any processor needs from other ones,

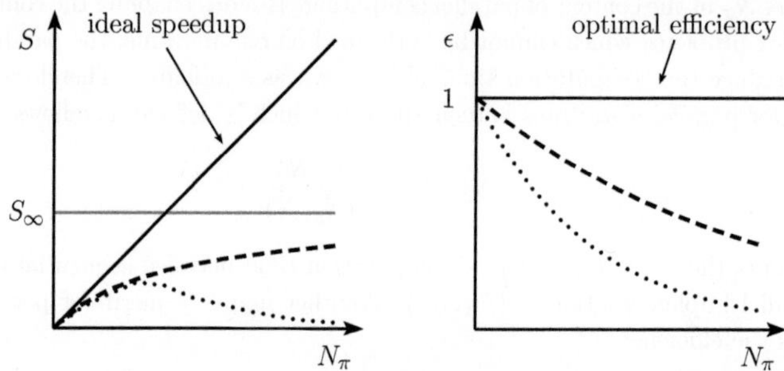

Figure 4.7: Speedup (left) and efficiency (right) w. r. t. the number of processors. Comparison between optimal and real behavior with (dotted line) and w/o (dashed line) considering the communication and synchronization task.

the higher the network activity gets. This leads even by using high speed network solutions to an overhead where less operations are performed as data exchanged. On the other side, a parallel computation can be as fast as the slowest involved processor. As a balanced implementation becomes more and more difficult by increased number of processors, the efficiency is further reduced. These two factors explain the maximum of the speedup in Figure 4.7 after which it decreases under real conditions. It is thus essential to determine the maximum number of processors depending on the problem size, the network activity and the degree of the problem to be balanced for optimal efficiency. Above this value, the parallelization task could be considered as a waste of resource.

Parallelization Strategy

The most time consuming parts of the MOR computation are

1. the LU decomposition which is done once before starting to build the Krylov space,

2. the forward and backward substitution along side with the modified Gram-Schmidt orthogonalization at each step of the Krylov's algorithm,

3. and the error estimation for which the computation at the 100 frequency samples is spread over the processors in presence.

It should be noted that the time needed to set up the system matrices and compute the transfer function (S- and Z-parameters) is negligible compared to the whole computation time.

An intuitive strategy to parallelize the MOR computation is the multi-point Padé approximation introduced in Section 3.4.2. The most suitable method for parallelization consists of setting a priori the expansion points and building the related Krylov spaces separately. The spaces are then merged and orthogonalized afterwards before being used for projection. As each processor would build a different space, this method offers an almost linear scalability assuming that the dimension of the overall space is equal or smaller than compared to the one resulting from the single point variant.

The structure used to test this method is a single wire modeled as a PEC body 5 cm above a PEC table (Figure 4.5). The two considered ports were placed at the ends of the wire and related to the table. The range of interest is 1 MHz to 4 GHz. The setups with different number of interpolation points are the following:

1. one interpolation point at 2 GHz,

2. three interpolation points at 0.6 GHz, 2 GHz and 3.3 GHz,

3. eight interpolation points equally spaced.

We investigated the convergence behavior of the reduction process of these setups by computing their relative error at dimensions 20, 40 and 80. The number of Krylov space vectors is equally distributed among the number of interpolation points. It should be noted that the number of vectors for each interpolation point is no more the same when the result of the division is not an integer. In that case, the rest of the division is spread over the first assigned processors.

The comparison of the relative error for a reduced system of dimension 20 is illustrated in Figure 4.11 and shows clearly the local convergence of the Padé approximation. In fact, the minima put in evidence that the transfer function is well approximated around the interpolation points. The larger convergence region for the single-point approximation comes from the fact that less vectors are generated for each expansion frequency in the multi-point variant.

Figure 4.12 shows the relative error in comparison for a reduction to dimension 40. It can be clearly seen that multi-point Padé approximation enables a more homogeneous convergence over the whole frequency range. For eight expansion frequencies,

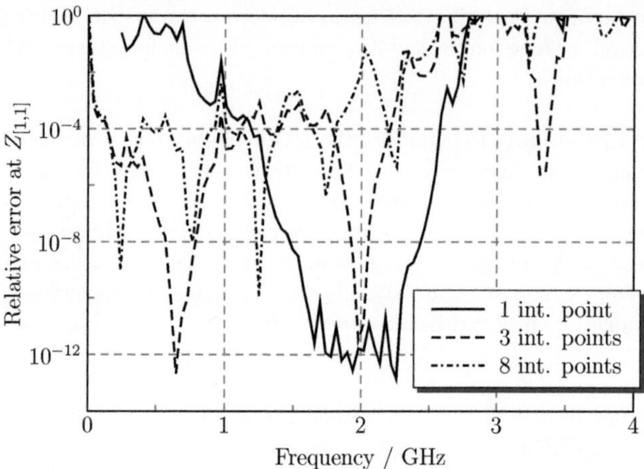

Figure 4.8: Comparison of the relative error at $Z_{[1,1]}$ of the wire model for the reduced system of dimension 20 with 1, 3 and 8 interpolation points.

the relative error is around 10^{-4} almost for the whole range. Considering 3 frequencies improves just slightly the convergence of the reduction process. The convergence radius of the frequencies lying above 2.5 GHz remains small as the amount of resonances there requires more vectors in order to be resolved.

Enlarging the projection space to 80 vectors degrade the multi-point Padé performance while the relative error drops below 10^{-8} for single point up to 3.75 GHz as can be seen in Figure 4.10. This confirms the fact the parallel generation of the Krylov spaces related to the different expansion frequencies results in numerical problems as mentioned in Section 3.4 [53]. This is basically due to the orthogonalization scheme. In fact, merging the Krylov spaces at the different frequencies at the end stage is less efficient than orthogonalizing the vectors gradually like in the modified Gram-Schmidt method (Algorithm 3.4.1). Therefore, we opted for the single-point variant in the parallelized version.

Implementation

The cluster on which the computations have to be run is hybrid. It consists of multi-computers which themselves have several processors. Therefore, we used MPI

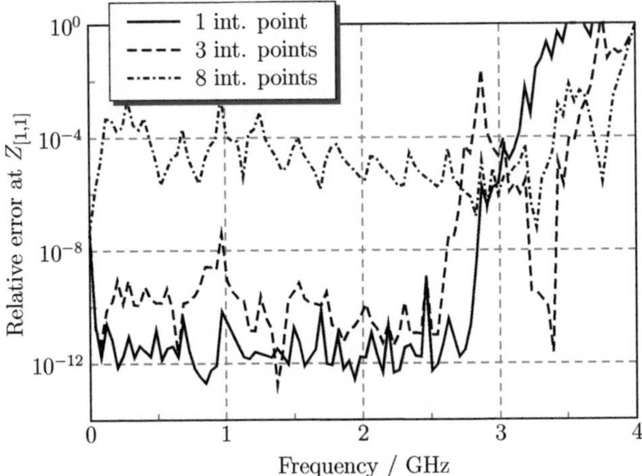

Figure 4.9: Comparison of the relative error at $Z_{[1,1]}$ of the wire model for the reduced system of dimension 40 with 1, 3 and 8 interpolation points.

as interface for the parallel implementation. In order to reduce the implementation efforts, we opted for the PETSc library [89] which includes an MPI interface. This library allows to handle objects like matrices or vectors in an easier way [90]. Furthermore, it relies on libraries for efficient algebra operations (e.g. LAPACK, BLAS [91]) and enables the integration of external solvers.

An analysis of the different steps of the code reveals that the LU decomposition is the most time consuming part of the computation. The influence of the generation of the Krylov space depends on the resonance behavior of the considered structure and the number of ports[5], so that its influence in the whole computation varies. However, it can be stated that in the structures considered in this work, this part takes less than 40% of the whole computation time. The computation of the Z- and S-parameters which takes less than 0.2% of the computation time can be parallelized in an efficient way, whereas the setup of the system matrices is sequential. It takes less than 1% of the computation time and thus, represents a negligible limiting factor for parallelization. Refer to [92] for more details on the implementation of the parallel MOR code.

---

[5]The more resonances a structure has, the more vectors are needed for the approximation.

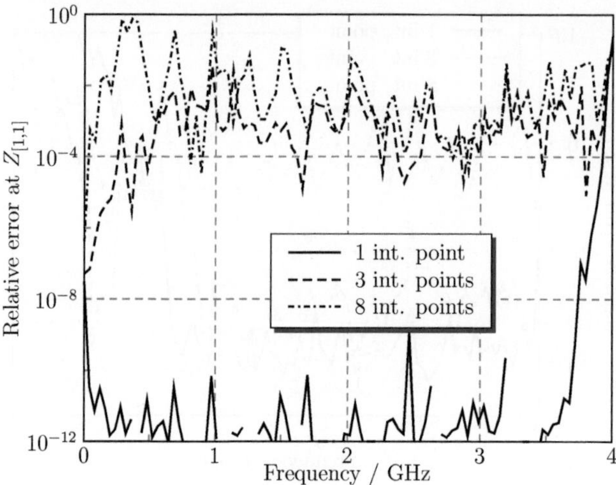

Figure 4.10: Comparison of the relative error at $Z_{[1,1]}$ of the wire model for the reduced system of dimension 80 with 1, 3 and 8 interpolation points.

The models of size 1.4×10^6, 3×10^6, and 8×10^6 DOFs, computed in this section are represented in Table 4.1.1:

1. Problem 1 is a converter with 6 ports.

2. Problem 2 is a 6 layered PCB model of a transmission control unit with 11 ports.

3. Problem 3 is the same model as problem 2 but finer meshed.

The performance of the MOR parallelization for these problems has been analyzed by means of speedup and efficiency comparisons. It should be stated that as the problems 1 and 2 could not be computed on one processor but at least on two, and four, respectively, the computation time has been extrapolated to one processor by multiplication[6].

---

[6]In fact, we assume a linear scalability from the minimum number of processors down to one as we could not get any computation time for comparison due to the high number of DOFs.

	Problem 1	Problem 2	Problem 3
# unknowns	1.4×10^6	3×10^6	8×10^6
# ports	7	11	11
# Krylov vectors	28	66	121

Table 4.1.1: Problems computed with the parallelized code. Problem 1 is a converter whereas problems 2 and 3 model a PCB for a transmission control unit with different mesh coarseness.

	2 proc.	4 proc.	8 proc.	16 proc.	32 proc.	48 proc.
LU	3459 s	1812 s	990 s	557 s	388 s	316 s
Krylov	352 s	253 s	234 s	195 s	168 s	101 s
Error	71 s	34 s	16 s	11 s	10 s	11 s

Table 4.1.2: Computation times of problem 1 with 1.4×10^6 DOFs.

Figures 4.11 and 4.12 show the speedup for the different computation stages (LU decomposition, Krylov space computation and error estimation) of problems 1 and 2 depending on the number of processors whereas the explicit computation times are presented in Tables 4.1.2 and 4.1.3. It can be clearly seen that the LU decomposition and the error estimation scale better than the Krylov space generation. In fact as the frequency samples are spread over all the processors while computing the error, a high efficiency can be achieved up to 16 processors (problem 1) and 32 processors (problem 2). After this limit, the overhead due to the communication becomes prejudicious. The LU decomposition scales better than the Krylov space generation because it is more intensive than the forward and backward substitution which consists solely on some matrix-vector multiplications.

Furthermore, Figure 4.13 which shows the efficiency of the whole computation for problems 1 and 2 reveal that the parallelization efficiency decreases from problem

	4 proc.	8 proc.	16 proc.	32 proc.	48 proc.
LU	2315 s	1279 s	730 s	518 s	426 s
Krylov	1051 s	1030 s	806 s	729 s	408 s
Error	297 s	147 s	71 s	52 s	50 s

Table 4.1.3: Computation times of problem 2 with 3×10^6 DOFs.

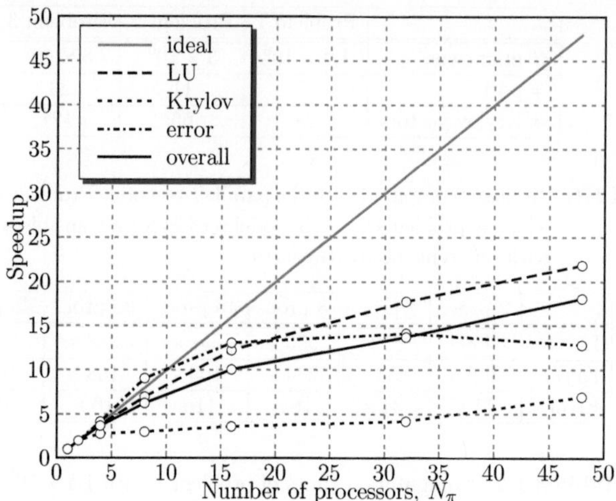

Figure 4.11: Speedup, S, of the overall computation and the different stages (LU decomposition, Krylov generation, and error estimation) w. r. t. the number of processors, N_π for problem 1 with 1.5×10^6 DOFs.

1 to 2. In fact, even though the LU decomposition scales better with increasing problem dimension, the higher proportion of the Krylov generation in the whole computation time due to the higher number of vectors in problem 2 (66 vs 28) worsens the whole efficiency. On 48 processors the efficiency is of 37.5% and 35%, respectively for problems 1 and 2, i.e. an acceleration factor of 18 and 16.8.

High performance computing as part of enabling the computation of complex structures allows for a faster computation as far as the hardware resources are available. For the 8×10^6 DOFs example, the LU decomposition could be computed in 1h30min with 48 processors whereas the computation on a 64GB machine with an iterative solver of MWS® in frequency domain could not be performed within a week. The computation times for 32 and 48 processors are given in Table 4.1.4.

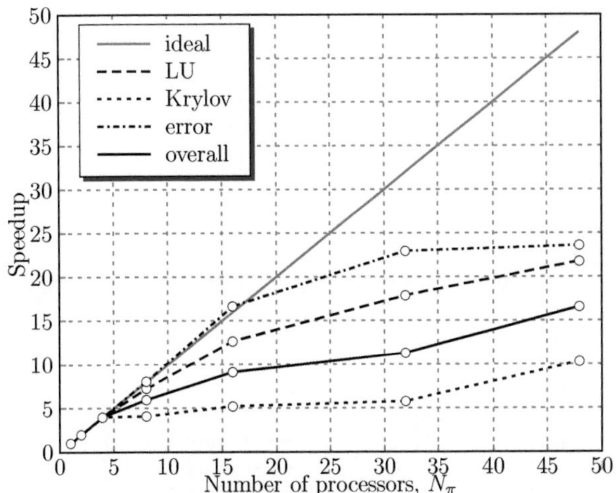

Figure 4.12: Speedup, S, of the overall computation and the different stages (LU decomposition, Krylov generation, and error estimation) w. r. t. the number of processors, N_π for problem 2 with 3×10^6 DOFs.

	32 proc.	48 proc.
LU	3806 s	2831 s
Krylov	4117 s	2775 s
Error	180 s	141 s

Table 4.1.4: Computation times of problem 3 with 8×10^6 DOFs on 32 and 48 processors.

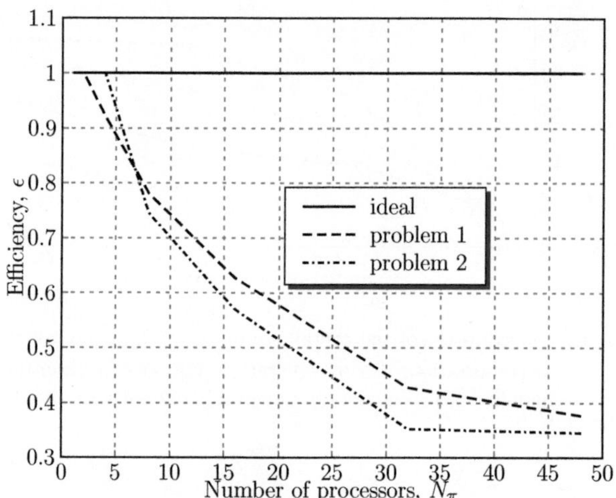

Figure 4.13: Comparison of the efficiency, ϵ, of the overall computation w. r. t. the number of processors, N_π between problem 1 with 1.5×10^6 DOFs and problem 2 with 3×10^6 DOFs.

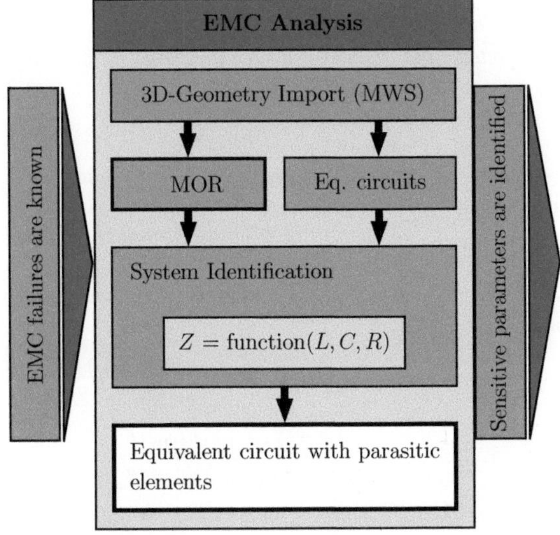

$$Z = \text{function}(L, C, R)$$

Figure 4.14: Overview of the block EMC analysis.

4.2 EMC Analysis

As already stated, MOR only provides a mathematical description of the behavior of the structure under test, whereas models which retrieve physical properties are indispensable for EMC analysis. In fact, such models which involve parasitic elements enable to understand the cause of the EMC effects detected in the previous step. Thus, knowing the sensitive parameters, improvement measures can be derived to retrieve an EMC optimized structure. Figure 4.14 shows in details how this block is constituted.

We used a semi-automatic method which assumes a set of possible equivalent circuits provided by the EMC engineer. These circuits consist of both functional and parasitic elements. The goal is to find the most realistic circuit and determine the values of its elements. For this purpose, the polynomial expressions from MOR and the MNA systems of the proposed circuits are fitted through optimization.

4.2.1 Polynomial Representation

The reduced models from PRIMA can be represented in rational polynomials

$$Z_{p[i,j]}(s) = \frac{\sum_{k=1}^{p-1} b_{k\text{MOR}}^{ij} s^k}{\sum_{l=1}^{p} a_{l\text{MOR}} s^l},\tag{4.2.1}$$

where the denominator is the same for all elements of the matrix $\mathbf{Z}$ as the poles refer to the whole system, and the numerator is specific to each $Z_{[i,j]}$. This expression builds the reference for the identification of the equivalent circuit.

Analogously, the MNA systems of the proposed circuits can be expressed as rational polynomials

$$Z_{\text{MNA}[i,j]}(s) = \frac{\sum_{k=1}^{m} b_{ik}\det(\mathbf{Y}_{kj})}{\det(\mathbf{Y})}\tag{4.2.2a}$$

$$= \frac{\sum_{k=1}^{p-1} b_{k\text{MNA}}^{ij} s^k}{\sum_{l=1}^{p} a_{l\text{MNA}} s^l},\tag{4.2.2b}$$

where $\mathbf{Y} = 1/s\mathbf{Y}_L + \mathbf{G} + s\mathbf{C}$ is the stamping matrix in (3.6.3) and $\mathbf{Y}_{kj}$ are defined as

$$\mathbf{Y}_{kj} = \begin{pmatrix} y_{1,1} & \cdots & y_{1,k-1} & b_{1,j} & y_{1_k+1} & \cdots & y_{1,n} \\ \vdots & \ddots & \vdots & \vdots & \vdots & \ddots & \vdots \\ y_{k,1} & \cdots & y_{k,k-1} & b_{k,j} & y_{k,k+1} & \cdots & b_{k,n} \\ \vdots & \ddots & \vdots & \vdots & \vdots & \ddots & \\ y_{n,1} & \cdots & y_{n,k-1} & b_{n,j} & y_{n,k+1} & \cdots & y_{n,n} \end{pmatrix},\tag{4.2.3}$$

where the kth column and k^{th} line of $\mathbf{Y}$ are replaced by the jth column of $\mathbf{B}$.

4.2.2 Identification

After having expressed both the reduced model and the MNA matrices of the different circuits as polynomials, a postprocessing step should be performed before identification. In fact, the frequency range of validity of reduced models obtained from the 3D discretization is broader than that of systems resulting from nodal analysis[7]. Therefore, after having set the frequency range $[f_{\min}, f_{\max}]$ in which the identification should be performed, only the zeros and poles smaller than $f_{\max}$ are considered

$$Z_{\text{MNA}[i,j]}(s) = \frac{\prod_{k=1}^{K}(s - s_k^{[ij]})}{\prod_{l=1}^{L}(s - s_l)}.\tag{4.2.4}$$

---

[7]The frequency range of 3D discretization goes typically up to GHz as the nodal analysis is no more valid for frequencies beyond 200 MHz for the structures under consideration in this work.

The following criteria are then used to identify the proper circuit with the values of its corresponding elements:

1. The first criterion is the **number of poles and zeros** L and K. The proposed circuits which do not have the same number of poles in the frequency range of validity should thus be eliminated.

2. The second criterion of identification is the **consistency of the computed values**. In fact, if the proposed circuit is wrong then the values retrieved by the optimization algorithm, at least for the functional elements, may go beyond an a-priori set range. This range should thus be treated as optimization constraints.

3. The last criterion is the **accuracy of the minimization function**. In fact, circuits which do not match the physical properties of the structure under test may not retrieve the transfer function with physical elements within the constraint.

The method would surely be more robust by defining the objective function in relation with poles and zeros to avoid some enhanced errors when dealing with badly conditioned polynomials. However, as the poles and zeros expressions resulting from the circuit analysis cannot be easily related to the right values from MOR[8], the function to be minimized is the mean of the relative errors of the coefficients in the formulations (4.2.1) and (4.2.2b)

$$f_{\text{opt}} = \frac{\sum_{i=1}^{m} |a_{i\text{MOR}} - a_{i\text{MNA}}|/|a_{i\text{MOR}}| + \sum_{i=1}^{n} |b_{i\text{MOR}} - b_{i\text{MNA}}|/|b_{i\text{MOR}}|}{m + n}, \quad (4.2.5)$$

where m and n is the number of all coefficients of the matrix $\mathbf{Z}$. After a first step of optimization, the poles and zeros can be better related to each other so that they could be used in a second step, if necessary.

4.3 EMC Optimization

EMC optimization consists of discrete measures, each consisting of different variants, which can be either considered separately or plugged together in an optimization setup. Therefore, an efficient computational method of the different variants is necessary. In the scope of this work, we considered variations of position and value of lumped elements and traces paths.

---

[8]The systems considered in this work have typically up to 10 poles and zeros.

4.3.1 Variation Computation

We assume that the variations are of local nature, so that the variations on the system matrix are of low rank. In this scope we consider variations of lumped elements, and of trace paths. As the matrix inversion in the reduction process is the main time consuming operation, we can use the so-called *matrix inversion lemma* for efficient computation [93].

Lemma 4.3.1 $\forall \mathbf{A} \in \mathbb{C}^{n \times n}$ *nonsingular,* $\mathbf{U}$ *and* $\mathbf{V} \in \mathbb{C}^{n \times m}$ *with* $m < n$, *if* $(\mathbf{I} + \mathbf{V}^H \mathbf{A}^{-1} \mathbf{U})$ *is nonsingular then the following holds*

$$(\mathbf{A} + \mathbf{U}\mathbf{V}^H)^{-1} = \mathbf{A}^{-1} - \mathbf{A}^{-1}\mathbf{U}(\mathbf{I} + \mathbf{V}^H \mathbf{A}^{-1}\mathbf{U})^{-1}\mathbf{V}^H \mathbf{A}^{-1} \qquad (4.3.1)$$

A lumped element placed on an edge induces a relation between current and voltage on the considered edge. Therefore, the curl-curl equation[9] introduced in (2.3.12) can be derived to

$$(j\omega)^2 \mathbf{M}_\varepsilon \, \hat{\mathbf{e}} + \mathbf{A}'_{CC}\, \hat{\mathbf{e}} = -j\omega(\hat{\hat{\mathbf{J}}} + \mathbf{Y}\,\hat{\mathbf{e}}), \qquad (4.3.2)$$

where $\mathbf{Y}$ is a diagonal matrix with elements $j\omega C$ and $\frac{1}{j\omega L}$, respectively, for capacitors and inductors as lumped elements. This leads to

$$(j\omega)^2(\mathbf{M}_\varepsilon + \mathbf{C}_{\text{cap}})\,\hat{\mathbf{e}} + (\mathbf{A}'_{CC} + \mathbf{L}_{\text{ind}}^{-1})\,\hat{\mathbf{e}} = -j\omega\,\hat{\hat{\mathbf{J}}}, \qquad (4.3.3)$$

where $\mathbf{C}_{\text{cap}}$ and $\mathbf{L}_{\text{ind}}$ are the diagonal matrices containing the values of considered capacitors and inductors, respectively.

Inductor Variations

The position and value variations of inductors imply a modification of the diagonal entries of $\mathbf{A}_{CC}$, as can be seen in (4.3.3). The modification has rank one, thus $\mathbf{U}$ and $\mathbf{V} \in \mathbb{R}^{N \times 1}$.

Capacitor Variations

Capacitor variations lead to a modification of $\mathbf{M}_\varepsilon^{-1/2}$ into

$$\hat{\mathbf{M}}_\varepsilon^{-1/2} = \mathbf{M}_\varepsilon'^{-1/2} + \Delta\mathbf{M}_\varepsilon^{-1/2}, \qquad (4.3.4)$$

where $\mathbf{M}_\varepsilon'^{-1/2}$ is nothing but $\mathbf{M}_\varepsilon^{-1/2}$ with the $m'_{\varepsilon[k,k]}$ being set to zero[10] and

$$\Delta m_{\varepsilon[k,k]}^{-1} = \frac{1}{m_{\varepsilon[k,k]} + C}. \qquad (4.3.5)$$

---

[9]We consider here the lossless case.
[10]We assume that the capacitor is placed at position k.

Thus, the variation of $\mathbf{A}_{CC}$ is derived to

$$\hat{\mathbf{M}}_\varepsilon^{-1/2}\mathbf{A}'_{CC}\hat{\mathbf{M}}_\varepsilon^{-1/2} = \tag{4.3.6}$$
$$\mathbf{A}_{CC} + \underbrace{\mathbf{M}_\varepsilon'^{-1/2}\mathbf{A}'_{CC}\Delta\mathbf{M}_\varepsilon^{-1/2} + \Delta\mathbf{M}_\varepsilon^{-1/2}\mathbf{A}'_{CC}\mathbf{M}_\varepsilon'^{-1/2} + \Delta\mathbf{M}_\varepsilon^{-1/2}\mathbf{A}'_{CC}\Delta\mathbf{M}_\varepsilon^{-1/2}}_{\Delta\mathbf{A}_{CC}=\mathbf{U}\mathbf{V}^T},$$

where $\mathbf{A}_{CC} = \mathbf{M}_\varepsilon'^{-1/2}\mathbf{A}'_{CC}\mathbf{M}_\varepsilon'^{-1/2}$ according to (2.3.11). This corresponds to a 2-rank modification of $\mathbf{A}_{CC}$ as the first and second terms relate to the kth column and k^{th} line respectively while the third term stands for the diagonal entry. The matrices $\mathbf{U} = [\mathbf{U}_1,\ldots,\mathbf{U}_r]$ and $\mathbf{V} = [\mathbf{V}_1,\ldots,\mathbf{V}_r] \in \mathbb{R}^{N\times 2r}$ with r being the number of capacitors can be defined as

$$\mathbf{U}_i = \begin{pmatrix} \dfrac{a_{CC[1,k]}\sqrt{m'_{\varepsilon[1,1]}}}{\sqrt{a_{CC[k,k]}m_{\varepsilon[k,k]}}} & -\dfrac{a_{CC[1,k]}\sqrt{m'_{\varepsilon[1,1]}}}{\sqrt{a_{CC[k,k]}m_{\varepsilon[k,k]}}} \\ \vdots & \vdots \\ \dfrac{2\sqrt{a_{CC[k,k]}m'_{\varepsilon[k,k]}}}{3\sqrt{\Delta m_{\varepsilon[k,k]}}} & -\dfrac{\sqrt{a_{CC[k,k]}m'_{\varepsilon[k,k]}}}{3\sqrt{\Delta m_{\varepsilon[k,k]}}} \\ \vdots & \vdots \\ \dfrac{a_{CC[N,k]}\sqrt{m'_{\varepsilon[N,N]}}}{\sqrt{a_{CC[k,k]}m_{\varepsilon[k,k]}}} & -\dfrac{a_{CC[N,k]}\sqrt{m'_{\varepsilon[N,N]}}}{\sqrt{a_{CC[k,k]}m_{\varepsilon[k,k]}}} \end{pmatrix} \tag{4.3.7a}$$

$$\mathbf{V}_i = \begin{pmatrix} \dfrac{3a_{CC[k,1]}\sqrt{m'_{\varepsilon[1,1]}}}{\sqrt{m_{\varepsilon[k,k]}}} & \dfrac{3a_{CC[k,1]}\sqrt{m'_{\varepsilon[1,1]}}}{\sqrt{m_{\varepsilon[k,k]}}} \\ \vdots & \vdots \\ \dfrac{2\sqrt{a_{CC[k,k]}m'_{\varepsilon[k,k]}}}{\sqrt{\Delta m_{\varepsilon[k,k]}}} & \dfrac{\sqrt{a_{CC[k,k]}m'_{\varepsilon[k,k]}}}{\sqrt{\Delta m_{\varepsilon[k,k]}}} \\ \vdots & \vdots \\ \dfrac{3a_{CC[k,N]}\sqrt{m'_{\varepsilon[N,N]}}}{\sqrt{m_{\varepsilon[k,k]}}} & \dfrac{3a_{CC[k,N]}\sqrt{m'_{\varepsilon[N,N]}}}{\sqrt{m_{\varepsilon[k,k]}}} \end{pmatrix} \tag{4.3.7b}$$

in order to guarantee the non-singularity of $(\mathbf{I} + \mathbf{V}^H\mathbf{A}^{-1}\mathbf{U})$ as required in Lemma 4.3.1. In fact, it is obvious that $\mathbf{U}_i$ and $\mathbf{V}_i$ correspond to the expression of $\Delta\mathbf{A}_{CC}$ in (4.3.6)

$$\mathbf{U}_i\mathbf{V}_i^T = \begin{cases} a_{CC[l,k]}\sqrt{m'_{\varepsilon[l,l]}}/\sqrt{\Delta m_{\varepsilon[k,k]}}, & \text{for } l=1,\ldots,N \text{ and } l \neq k \\ a_{CC[k,l]}\sqrt{m'_{\varepsilon[l,l]}}/\sqrt{\Delta m_{\varepsilon[k,k]}}, & \text{for } l=1,\ldots,N \text{ and } l \neq k \\ a_{CC[k,k]}m'_{\varepsilon[k,k]}/\Delta m_{\varepsilon[k,k]} \\ 0, & \text{else.} \end{cases} \tag{4.3.8}$$

Furthermore, the non-singularity of $(\mathbf{I} + \mathbf{V}^T \mathbf{A}^{-1} \mathbf{U})$ can be proved by showing that $\mathbf{V}_i^T \mathbf{U}_i$ is not singular[11]. Obviously, the term

$$\mathbf{V}_i^T \mathbf{U}_i = \begin{pmatrix} vu_1 & -vu_2 \\ vu_2 & -vu_3 \end{pmatrix}, \tag{4.3.9}$$

with

$$vu_1 = \sum_{l \neq k} \frac{3a_{CC[l,k]}^2 m'_{\varepsilon[l,l]}}{a_{CC[k,k]} m'_{\varepsilon[k,k]}} + \frac{4a_{CC[k,k]} m'_{\varepsilon[k,k]}}{3\Delta m_{\varepsilon[k,k]}}, \tag{4.3.10a}$$

$$vu_2 = vu_1 - \frac{2a_{CC[k,k]} m'_{\varepsilon[k,k]}}{3\Delta m_{\varepsilon[k,k]}}, \tag{4.3.10b}$$

$$vu_3 = vu_2 - \frac{a_{CC[k,k]} m'_{\varepsilon[k,k]}}{3\Delta m_{\varepsilon[k,k]}}, \tag{4.3.10c}$$

has similar entries up to the factor $\frac{a_{CC[k,k]} m'_{\varepsilon[k,k]}}{3\Delta m_{\varepsilon[k,k]}}$. Thus the matrix $\mathbf{V}_i^T \mathbf{U}_i$ guarantees a good conditioned application of the inversion lemma. It should be noted that we considered the fact that $\mathbf{A}_{CC}$ is symmetric in (4.3.10c).

Trace Path Variations

We interpret any trace as a PEC body. They are characterized by their vanishing electric field. This is modeled by enforcing the relative permittivity to tend to infinity or its reciprocal to be zero inside and on the surface of PEC materials. Again, we obtain the following matrices:

$$\mathbf{U}_i = \begin{pmatrix} -\frac{a_{CC[1,k]}}{\sqrt{a_{CC[k,k]}}} & \frac{a_{CC[1,k]}}{\sqrt{a_{CC[k,k]}}} \\ \vdots & \vdots \\ -\frac{2}{3}\sqrt{a_{CC[k,k]}} & \frac{1}{3}\sqrt{a_{CC[k,k]}} \\ \vdots & \vdots \\ -\frac{a_{CC[N,k]}}{\sqrt{a_{CC[k,k]}}} & \frac{a_{CC[N,k]}}{\sqrt{a_{CC[k,k]}}} \end{pmatrix}, \quad \mathbf{V}_i = \begin{pmatrix} \frac{3a_{CC[k,1]}}{\sqrt{a_{CC[k,k]}}} & \frac{3a_{CC[k,1]}}{\sqrt{a_{CC[k,k]}}} \\ \vdots & \vdots \\ 2\sqrt{a_{CC[k,k]}} & \sqrt{a_{CC[k,k]}} \\ \vdots & \vdots \\ \frac{3a_{CC[k,N]}}{\sqrt{a_{CC[k,k]}}} & \frac{3a_{CC[N,k]}}{\sqrt{a_{CC[k,k]}}} \end{pmatrix} \tag{4.3.11}$$

which can also be used for the *matrix inversion lemma*. The entries of the modification are derived to

$$\mathbf{U}_i \mathbf{V}_i^T = \begin{cases} -a_{CC[l,k]}, & \text{for } l = 1, \ldots, N \text{ and } l \neq k \\ -a_{CC[k,l]}, & \text{for } l = 1, \ldots, N \text{ and } l \neq k \\ 0, & \text{else.} \end{cases} \tag{4.3.12}$$

---

[11]Note that we use the subscript T instead of H because the matrices $\mathbf{U}$ and $\mathbf{V} \in \mathbb{R}$ in our setup.

which eliminate the corresponding entries in $\mathbf{A}_{CC}$. The term is also non-singular as

$$\mathbf{V}_i^T \mathbf{U}_i = \begin{pmatrix} vu_1 & -vu_2 \\ vu_2 & -vu_3 \end{pmatrix}, \tag{4.3.13}$$

with

$$vu_1 = \sum_{l \neq k} \frac{3a_{CC[l,k]}^2}{a_{CC[k,k]}} + \frac{4}{3}a_{CC[k,k]}, \tag{4.3.14a}$$

$$vu_2 = vu_1 - \frac{2}{3}a_{CC[k,k]}, \tag{4.3.14b}$$

$$vu_3 = vu_2 - \frac{1}{3}a_{CC[k,k]}, \tag{4.3.14c}$$

$$\tag{4.3.14d}$$

has entries of same order like in (4.3.9).

It is more difficult in this case to set the positions of $\mathbf{M}_\varepsilon$ which should be changed. In order to simplify this process we defined a way to derive the positions with just the minimal and maximal node as input values (Fig.4.15).

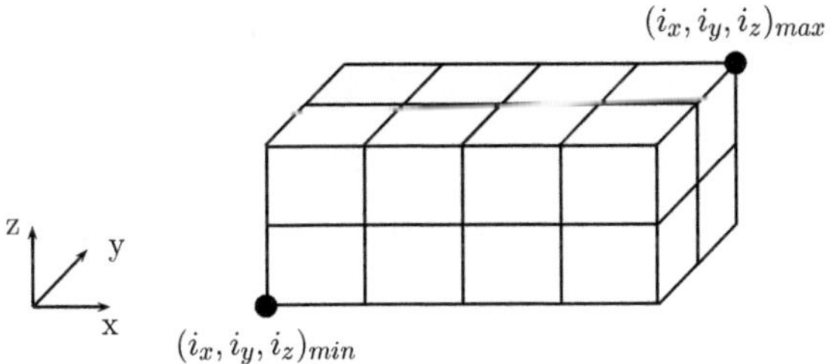

Figure 4.15: Meshed PEC trace with the 2 identification points

This method is initially limited to pieces of parallelograms as further improvements are required to consider general structures. Assuming

$$m_x = i_{x\text{max}} - i_{x\text{min}} \tag{4.3.15a}$$

$$m_y = i_{y\text{max}} - i_{y\text{min}} \tag{4.3.15b}$$

$$m_z = i_{z\text{max}} - i_{z\text{min}}, \tag{4.3.15c}$$

it can be proved that the modification rank is

$$r = 2(m_x + m_y + m_z + 2(m_x m_y + m_x m_z + m_y m_z) + 3m_x m_y m_z). \qquad (4.3.16)$$

4.3.2 Optimization Flow

Although it has been proved to be exact, the method through the matrix inversion lemma fails unfortunately in practice to be more efficient than the intuitive one. In fact, we use PARDISO [94][12] to solve the equation at each Krylov iteration step. It should be stated that SUPERLU was used for parallelization because PARDISO is applicable only on SMMs and not on DMMs. PARDISO performs two main tasks:

1. numerical LU factorization,

2. forward and backward substitution.

Assuming t_1 and t_2 as the times required to perform tasks 1 and 2, the condition for efficiency can be given as follows

$$(t_1 + p \cdot t_2)n_{\text{opt}} > t_1 + n_{\text{opt}} \cdot p \cdot r \cdot t_2, \qquad (4.3.17)$$

where p is the dimension of the Krylov space, r the rank of modification, and n_{opt}, the number of optimization iterations. This expression can be derived to

$$t_1(n_{\text{opt}} - 1) > n_{\text{opt}} \cdot p \cdot t_2(r - 1) \Rightarrow$$
$$\gamma = \frac{t_1}{t_2} > p \cdot r, \qquad (4.3.18)$$

by assuming $n_{\text{opt}} \gg 1$ and $r \gg 1$. In practice, the ratio between the two tasks of PARDISO is of order $\mathcal{O}(10^2)$ as illustrated in Figure 4.16, while the product $p \cdot r$ is at least 10 times greater for complex systems. Therefore, this method can not be used for our purposes.

However, the ability to modify the system matrices after each optimization iteration is advantageous. In fact, our optimization flow, as illustrated in Figure 4.17, requires the mesh data only once from MWS® whereas the main optimization task is performed as a stand alone process.

After having identified the sensitive EMC measures and set the range for position and value or length of trace paths and/or lumped elements, the optimization algorithm[13] can be started. It is worth noting that the traces which should be modified

---

[12]The best currently available LU solver for our applications [95]
[13]This could be the optimization tool OptiSlang [96].

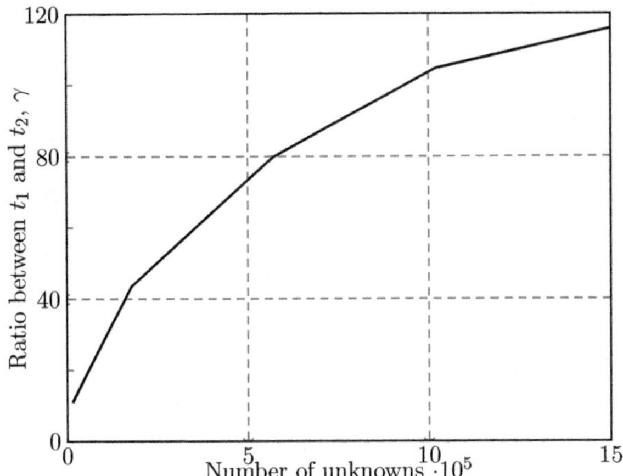

Figure 4.16: Ratio of computation time between the two PARDISO tasks (numerical LU factorization and forward and backward substitution) vs. number of unknowns.

are not considered in the meshing process. However, the mesh should be fine enough at the regions where they should be placed in order to perform the modifications in the optimization process with better accuracy. The objective value can be computed by considering either the transfer function or currents and voltages after a circuit simulation[14]. The optimal parameters can then be retrieved if a given stop criterion has been reached.

Even though the efficiency could not be improved by using the *matrix inversion lemma*, the possibility to modify the system matrices enables an automatic optimization without requiring new mesh information after each optimization step. Furthermore, it allows the flexibility to remove or introduce some traces within the optimization process which cannot be achieved easily in the MWS® environment[15]. Again as can be seen in Figure 4.18 which resumes the task performed in this section, MOR plays a central role by enabling a faster optimization.

---

[14]Generally some resonances should removed or shifted.

[15]One would have to write some complex macros in VisualBasic in order to make it available.

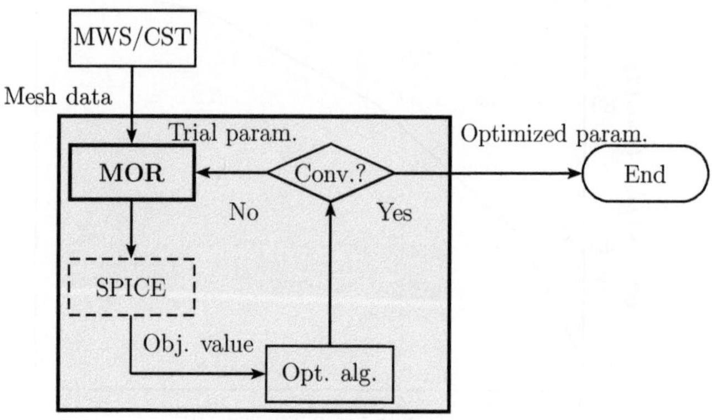

Figure 4.17: Optimization workflow.

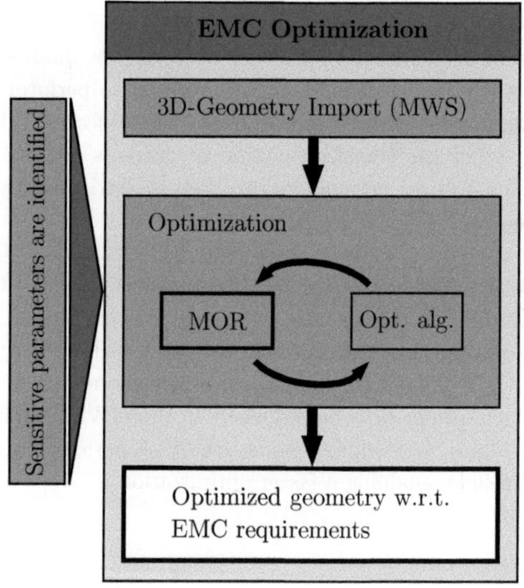

Figure 4.18: Overview of the block EMC optimization.

5 Matrix Compression

In this chapter, the Kronecker decomposition is presented as an efficient way to compress matrices arising from FIT systems and thus allow computations with an efficient memory management. After a short introduction, a brief overview on $\mathcal{H}$-matrices which have been also introduced for matrix compression is given for comparison. Then, the Kronecker decomposition is described where particularly the approximation of the inverse and the so-called Kronecker-Krylov are addressed.

5.1 Introduction

In order to cope with the challenge of growing memory and complexity requirements, we distinguish two strategies:

- One could try to adapt the hardware to the existing software by solving high dimensional problems on clusters through parallelization in order to cope with the storage demand.

- Another goal is to develop software which could enable the computation of such problems on machines with moderate memory capacity.

For this last option, several methods of matrix compression which allow an almost linear complexity[1] have already been introduced and applied in the past to represent full matrices arising from integral operators (BEM):

- The multipole method [97]

- Panel-clustering [98]

- $\mathcal{H}$-Matrices [99]

- ACA Matrices [100, 101]

Recently a new method, the so-called *Kronecker-decomposition*, has been introduced and successfully applied on sparse matrices arising from volume discretization

---

[1] almost linear means $\mathcal{O}(nlog^k n)$ for $k \in \mathbb{N}$

techniques [102]. The next section presents the theory of $\mathcal{H}$-Matrices in order to show the difference between the class of methods enumerated above and the main concern of this chapter: *Kronecker-decomposition*. It should be stated that we will only present the potential of the *Kronecker-decomposition* in the computation of FIT problems as the implementation of this method goes beyond the focus of this work.

5.2 $\mathcal{H}$-Matrices

$\mathcal{H}$-matrices where $\mathcal{H}$ stands for "hierarchical" were first introduced by Hackbusch in 1999 [99]. This method close to *panel clustering* sets a class of matrices which represent full matrices, e.g. matrices arising from integral operators, in a data-sparse way.

Let a matrix $\mathbf{A}$ be defined as an $\mathcal{R}_k$-matrix if $rank(\mathbf{A}) \leq k$ and let us restrict our considerations for the following to the case $k = 1$. The simplest description of an $n \times n$ $\mathcal{H}$-matrix $\mathbf{A}$ with $n = 2^p$ $(p > 0)$ can be given by the constraint that $\mathbf{A}$ has the block structure

$$\mathbf{A} = \begin{pmatrix} \mathbf{A}_{11} & \mathbf{A}_{12} \\ \mathbf{A}_{21} & \mathbf{A}_{22} \end{pmatrix} \qquad (5.2.1)$$

with the $\frac{n}{2} \times \frac{n}{2}$ $\mathcal{H}$-matrices $\mathbf{A}_{ii}$ and the $\mathcal{R}_1$-matrices $\mathbf{A}_{12}$, $\mathbf{A}_{21}$. This procedure is performed down to 1×1 $\mathcal{H}$-matrices which are 1×1 matrices. In other words, a $\mathcal{H}$-matrix compresses a full matrix in a non-tensor partitioned block of low-rank matrices which are decomposed to factors by singular value decomposition (SVD)-like approximations. The fact that the partitioning is no more tensor-like for $p > 1$ can be seen in Figure 5.1.

Any $n \times m$-matrix $\mathbf{A}$ of rank ≤ 1 can be written in the form

$$\mathbf{A} = a^H b \qquad (5.2.2)$$

with $a \in \mathbb{C}^n$, $b \in \mathbb{C}^m$ and b^H being the Hermitian transpose of b. The amount of storage is $n + m$, which is $\mathcal{O}(n + m)$ for $\mathcal{R}_k$-matrices.

The sum of two $\mathcal{R}_1$-matrices has in general rank 2. The SVD helps here to approximate this sum through an $\mathcal{R}_1$-matrix ($\mathcal{R}_1$-addition). Let $\mathbf{A}$ be a $n \times m$-matrix of rank k. The matrix

$$\mathbf{A}' = \mathbf{U}\mathbf{D}'\mathbf{V} \text{ with } \mathbf{D}' := \text{diag}\{d_1, \ldots, d_{k'}, 0, \ldots, 0\} \qquad (5.2.3)$$

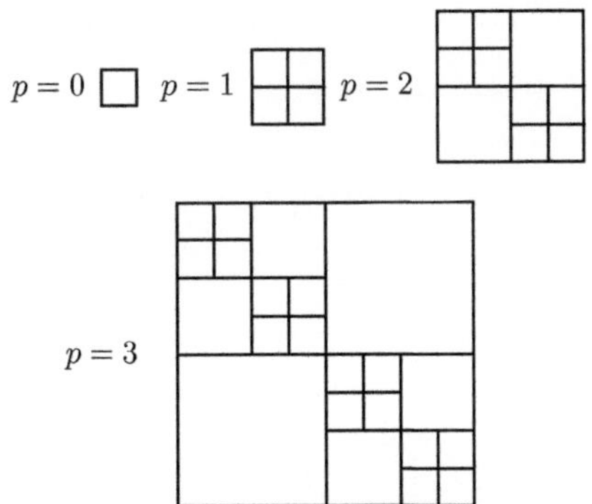

Figure 5.1: Block partitioning following the $\mathcal{H}$-matrix theory

where $k' \in [1, k]$ is the rank k' approximation of $\mathbf{A}$ with smallest Frobenius norm[2].

The memory requirement for any $n \times n$ $\mathcal{H}$-matrix with $n = 2^p$ is

$$\begin{aligned}
N_{mem} &= 2N_{\mathcal{R}_1}(p-1) + 2N_{mem}(p-1) & (5.2.4)\\
&= (1 + 2\log_2 n)n & (5.2.5)
\end{aligned}$$

The complexity of the addition of two $n \times n$ $\mathcal{H}$-matrices is $\mathcal{O}(18n\log_2 n)$ and the matrix-vector multiplication requires $4n\log_2 n$ operations.

5.3 Kronecker-Decomposition

This method is based on the multilinear decomposition well known in statistics [103, 104]. It has been first introduced by Ibraghimow [102] for compression of matrices arising from integral operators on cartesian grids.

5.3.1 Kronecker-Product

The Kronecker-product and its properties build the basis of this decomposition. Let $\mathbf{B} \in \mathbb{C}^{n_1 \times n_1}$ and $\mathbf{C} \in \mathbb{C}^{n_2 \times n_2}$, then the Kronecker product $\otimes$ is defined like

$$\mathbf{B} \otimes \mathbf{C} = \begin{pmatrix} b_{[1,1]}\mathbf{C} & \cdots & b_{[1,n_1]}\mathbf{C} \\ \vdots & \ddots & \vdots \\ b_{[n_1,1]}\mathbf{C} & \cdots & b_{[n_1,n_1]}\mathbf{C} \end{pmatrix} \in \mathbb{C}^{n_1 n_2 \times n_1 n_2}. \qquad (5.3.1)$$

---

[2]$\|\mathbf{A}\|_F = \sqrt{\sum_{i,j} a_{[i,j]}^2}$

Some of the basic properties of the Kronecker-product are [102]

$$
\begin{aligned}
(B \otimes C)^* &= (B^* \otimes C^*) && \text{(5.3.2)} \\
(B \otimes C)(D \otimes F) &= (BD \otimes CF) && \text{(5.3.3)} \\
B \otimes (C \otimes D) &= (B \otimes C) \otimes D && \text{(5.3.4)}
\end{aligned}
$$

5.3.2 Decomposition

In the following, we assume $N = n_1 \times \cdots \times n_d$ to be the dimension of the matrix $\mathbf{A}$. In our *curl-curl* case, $d = 4$ where n_1, n_2 and n_3 represent the number of grid points for each direction (x, y, and z) and $n_4 = 3$ stands for the vectorial form.

A Kronecker-decomposition of $\mathbf{A}$ with Kronecker rank ($\mathcal{K}$-rank$(\mathbf{A})$) r can be given as

$$
\mathbf{A} = \sum_{l=1}^{r} \mathbf{A}_l^{(1)} \otimes \cdots \otimes \mathbf{A}_l^{(d)}, \tag{5.3.5}
$$

$$
\mathbf{A}_l^{(p)} = [a_{i_p j_p l}]_{i_p, j_p = 1}^{n_p, n_p}.
$$

The compression factor of this method is high ($\mathcal{O}(drN^{2/d})$ [105]) and even sublinear for 3D problems. It should be noted that this requires $r < \min\{n_1, \ldots, n_d\}$.

The product of two Kronecker-matrices of rank r_A and r_B

$$
\mathbf{AB} = \sum_{l,l'=1}^{r_A, r_B} \left(\mathbf{A}_l^{(1)} \mathbf{B}_{l'}^{(1)} \otimes \cdots \otimes \mathbf{A}_l^{(d)} \mathbf{B}_{l'}^{(d)} \right) \tag{5.3.6}
$$

has an arithmetic complexity of $\mathcal{O}(dr_A r_B N^{3/d})$ which is linear for $d = 3$ and $\mathcal{K}$-$rank(\mathbf{AB}) = r_A r_B$.

Analogously, a vector with Kronecker rank r_v can be represented as follows:

$$
\mathbf{v} = \sum_{l=1}^{r_v} \mathbf{v}_l^{(1)} \otimes \cdots \otimes \mathbf{v}_l^{(d)} \tag{5.3.7}
$$

$$
\mathbf{v}_l^{(p)} = [v_{i_p l}]_{i_p}^{n_p}.
$$

The Kronecker vector-matrix product

$$
\mathbf{Av} = \sum_{l,l'=1}^{r_A, r_v} \left(\mathbf{A}_l^{(1)} \mathbf{v}_{l'}^{(1)} \otimes \cdots \otimes \mathbf{A}_l^{(d)} \mathbf{v}_{l'}^{(d)} \right) \tag{5.3.8}
$$

has a sublinear complexity of $\mathcal{O}(dr_A r_v N^{3/d})$ and $\mathcal{K}\text{-}rank(\mathbf{A}v) = r_A r_v$.

A more general decomposition schema, the so-called $Kronecker - Tucker$ decomposition, has been proven in [106] to be more efficient than (5.3.6). In the following, we will consider the first 3 dimensions without loss of generality as the extension to the fourth is trivial.

The matrices resulting from the volume discretization can be interpreted as 3D-tensors:

$$\mathbf{A}_1 = [a^{(1)}_{[s_1,i_1]}] \in \mathbb{C}^{m_2 m_3 \times m_1}, a^{(1)}_{[s_1,i_1]} = a_{[i_1,i_2,i_3]}, s_1 = i_2 + (i_3 - 1)m_2, \qquad (5.3.9a)$$

$$\mathbf{A}_2 = [a^{(2)}_{[s_2,i_2]}] \in \mathbb{C}^{m_3 m_1 \times m_2}, a^{(2)}_{[s_2,i_2]} = a_{[i_1,i_2,i_3]}, s_2 = i_3 + (i_1 - 1)m_3, \qquad (5.3.9b)$$

$$\mathbf{A}_3 = [a^{(3)}_{[s_3,i_3]}] \in \mathbb{C}^{m_1 m_2 \times m_3}, a^{(3)}_{[s_3,i_3]} = a_{[i_1,i_2,i_3]}, s_3 = i_1 + (i_2 - 1)m_1, \qquad (5.3.9c)$$

with $m_p = n_p^2$. The decomposition is then given as

$$a_{[i_1,i_2,i_3]} = \sum_{i_1',i_2',i_3'=1}^{r_1,r_2,r_3} b^{(1)}_{[i_1,i_1']} b^{(2)}_{[i_2,i_2']} b^{(3)}_{[i_3,i_3']} g_{[i_1',i_2',i_3']}, \qquad (5.3.10)$$

where $\forall k = 1, 2, 3$, $\mathbf{B}_k = [b_{[i_k,i_k']}]^{m_k,r}_{[i_k,i_k'=1]}$ are the orthonormal Q-factors of the QR-decompositions of $\mathbf{A}_k^T$ with rank r_k. The coefficients $g_{i_1',i_2',i_3'}$ can be computed the following way:

$$g_{[i_1',i_2',i_3']} = \sum_{i_1,i_2,i_3=1}^{m_1,m_2,m_3} b^{(1)}_{[i_1,i_1']} b^{(2)}_{[i_2,i_2']} b^{(3)}_{[i_3,i_3']} a_{[i_1,i_2,i_3]}. \qquad (5.3.11)$$

The $Kronecker\text{-}Tucker$-decomposition has an arithmetical complexity of $\mathcal{O}(m_1 + m_2 + m_3)$ and a memory requirement of $\mathcal{O}(\sum_{p=1}^{3} r_p n_p^2 + \prod_{p=1}^{3} r_p)$ [106]. Table 5.3.1 shows the potential of this method on an example with 21×10^6 DOFs.

While this method may lead to high compression rates for dense matrices, its efficiency for sparse matrices resulting from FIT discretization which have already a linear complexity is low. Therefore, its efficiency can be improved by applying it to approximate the inverse of the FIT matrices which are dense. This has been made possible in [105] by means of Newton iteration which are defined as [107]

$$\mathbf{P}_{i+1} = 2\mathbf{P}_i - \mathbf{P}_i \mathbf{A} \mathbf{P}_i \qquad (5.3.12a)$$

$$\text{or } \mathbf{P}_{i+1} = \mathbf{P}_i(3\mathbf{I} - \mathbf{A}\mathbf{P}_i(3\mathbf{I} - \mathbf{A}\mathbf{P}_i)), \qquad (5.3.12b)$$

	Speicherplatz
System matrix	$7GB$
Preconditioner (ILU)	$n \times 7\text{GB}(n > 5)$
Decomposition with rank 50	0.6GB
Decomposition with rank 100	1.1GB
Decomposition with rank 150	1.7GB

Table 5.3.1: Potential of the Kronecker-decomposition compared to an incomplete LU decomposition (ILU)

where the start iteration value could be a scaled unity matrix. As the rank of the matrices $\mathbf{P}_i$ may grow tremendously, a low-rank approximation $\widetilde{\mathbf{P}}_i$ is computed by minimizing the error

$$\|\mathbf{P}_i - \widetilde{\mathbf{P}}_i\|_F < \varepsilon, \qquad (5.3.13)$$

where $\|\cdot\|_F$ is the Frobenius norm and ε is a given threshold value. It has been proved in [105] that this method leads to approximated inverses with a very low Kronecker rank ($r \simeq 10$) for matrices of order of 10^6 resulting from Laplace operators.

5.3.3 Kronecker-Krylov

Besides the inversion of the matrix, the storage of the Krylov space represents the highest limit for the model order reduction process. In [106], Kronecker and Krylov methods were combined to solve linear systems of equations or eigenvalue problems.

This so-called Kronecker-Krylov method has been extended to reduce the memory requirement and computational complexity of the Padé approximation. It consists of approximating the vectors of the Krylov space $\mathbf{V}_p = span\{\mathbf{B}, \mathbf{AB}, \ldots, \mathbf{A}^{p-1}\mathbf{B}\}$ by vectors of $\mathcal{K}$-rank equal to 1. The generation of the Krylov space is then modified to [106]:

where α and $\mathbf{v}_j$ are normalized Kronecker vectors of $\mathcal{K}$-rank equal to 1. The Cholesky decomposition of $\mathbf{V}_i$, $\mathbf{V}_i^*\mathbf{V}_i = \mathbf{L}_i^*\mathbf{L}_i$ yields the unitary matrix $\mathbf{Q}_i = \mathbf{V}_i\mathbf{L}_i^{-*}$ [3] which is similar to the Krylov space. This algorithm is similar to the partial realization presented in Section 3.4.2. It has a linear complexity ($\mathcal{O}(iN + i^3)$) and a sublinear memory requirement for 3D problems ($\mathcal{O}(iN^{1/3})$).

Table 5.3.2 shows the efficiency of this method compared to the already introduced partial realization with respect to memory requirement.

---

[3] $\mathbf{Q}_i^*\mathbf{Q}_i = \mathbf{L}_i^{-1}\mathbf{V}_i^*\mathbf{V}_i\mathbf{L}_i^{-*}$

Algorithm 5.3.1 Kronecker-Krylov algorithm

$\mathbf{v}_1 = \mathbf{b}/\|\mathbf{b}\|_2, \mathbf{V}_1 = [\mathbf{v}_1]$
for $i = 2$ to m **do**
  $\mathbf{v}_i = \mathbf{b}_i/\|\mathbf{b}_i\|_2, \mathbf{V}_i = [\mathbf{V}_{i-1}, \mathbf{v}_i]$
end for
for $j = m + 1$ to p **do**
  $\mathbf{r}_j = \mathbf{A}\mathbf{V}_{j-1}\mathbf{t}_j, \ \mathbf{t}_j \in \mathbb{C}^{j-1}$
  $\min\|\mathbf{r}_j - \mathbf{V}_{j-1}\alpha - \beta\mathbf{v}_j\|_2; \ \|\mathbf{v}_j\|_2 = \|\alpha\|_2 = 1$
  $\mathbf{V}_j = [\mathbf{V}_{j-1}, \mathbf{v}_j]$
end for

	Partial realization	Kronecker-Krylov
100 vectors	16 GB	700 MB
500 vectors	80 GB	3.5 GB
1000 vectors	160 GB	7 GB

Table 5.3.2: Potential of the Kronecker-Krylov method as order reduction method compared to partial realization

5.4 Conclusion

The concept of $\mathcal{H}$-matrices and Kronecker-decomposition and their application areas have been presented. $\mathcal{H}$-matrices are well suited for dense matrices, but the compression efficiency (almost linear) is poor for sparse matrices. However, $\mathcal{H}$-matrices can be applied to store the inverse of sparse matrices as they are in general dense. Nowadays, they are often used to deal with matrices arising from the Boundary Element Method (BEM).

The Kronecker-decomposition is not only for dense matrices of great benefit as its sublinear storage requirement is also efficient for sparse matrices. This method can also be used to approximate the inverse of matrices. Though, it can only be applied on rectangular grids (FIT systems) and is suited for problems with low Kronecker-rank i.e. with a weak dependance among the 3 dimensions in space. In addition, this decomposition is tensor-like whereas the partitioning of $\mathcal{H}$-matrices is not. Furthermore, the fact that the generation of Krylov vectors could be performed more efficiently with regards to the memory requirements makes this method promising for MOR computations on complex systems with more than 10^7 DOFs .

6 Results

As the focus of this work is the application of numerical methods for efficient EMC engineering, some real-life examples have been picked and analyzed with the workflow introduced in Sections 1 and 4:

1. *ESP model: A setup which aims to analyze the common mode induced effects has been measured in our department and simulated in this work.*

2. *DC/DC converter: The converter which has also been measured in our department has been optimized by means of the method introduced in Section 4.*

6.1 ESP Model

6.1.1 Model Description

The main goal of this analysis is to understand the common mode effects occurring in the measurement setup illustrated in Figure 6.1. It consists of

- a PCB modeled as a two-layer plate for simplicity placed 5 cm above a measurement table,

- two 20 cm wires attached at one end to the PCB ground and at the other to two ports which are used to monitor the voltages at that end,

- two traces (long and short) are placed on the signal layer of the PCB.

The computational domain is enclosed with PMC boundaries except of the lower part (table) which is bounded by a PEC body. We do not consider any losses as all metal bodies are modeled as PEC. The discretization with MWS yields $650,000$ unknowns.

In the measurement setup, the trace[1] is excited by an FPGA[2] signal. This signal is modeled as a sequence of pulses with a width of 20 ns and a period of 1 μs. The

---

[1]The trace is modeled by the longest path in Figure 6.1.

[2]Field programmable gate array

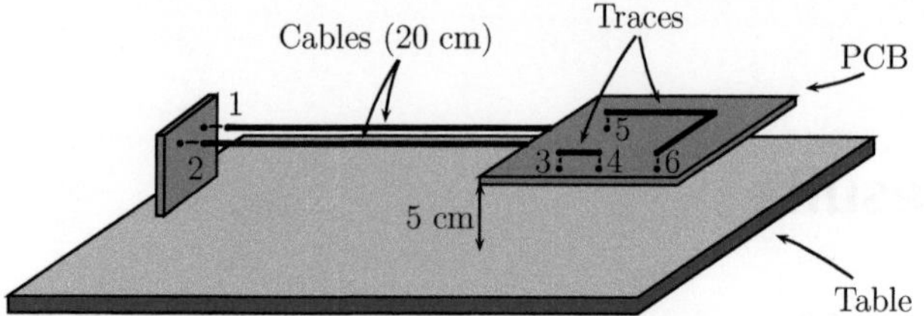

Figure 6.1: Measurement setup model of ESP attached to wires.

distortions caused by common mode effects are then measured at the end of one of the two cables. The six ports in the model are defined as follows:

- ports 1 and 2 are placed between one end of the wire and ground,

- ports 3 and 5 are used for excitation of the traces,

- and ports 4 and 6 represent the loads of the traces.

In order to consider the FPGA signal, a circuit simulation into which the 6-port model is plugged (Figure 6.2) is performed. Port 2 is shorted as the second wire is directly connected to the ground in measurements. As we first consider the longer trace, ports 3 and 4 are terminated with 50 Ω and 150 Ω loads as the FPGA signal is applied at port 5. Meanwhile, port 6 is connected to a 150 Ω load.

6.1.2 Computation

As already stated, the EMC prediction in this example comprises two steps:

1. The first one consists of a 3D field simulation after which the transfer function is computed and a macromodel is extracted.

2. The second step consists of a circuit simulation in which the voltage at port 1 is retrieved.

3D Field Simulation

The frequency range in which the computation is performed is $[1\ \mathrm{MHz}, \ldots, 1\ \mathrm{GHz}]$. The expansion frequency of the reduction scheme is set at 0.6 GHz. Furthermore, the minimum dimension of the reduced model is 36 after what the error estimation is performed with an increment of 12 vectors.

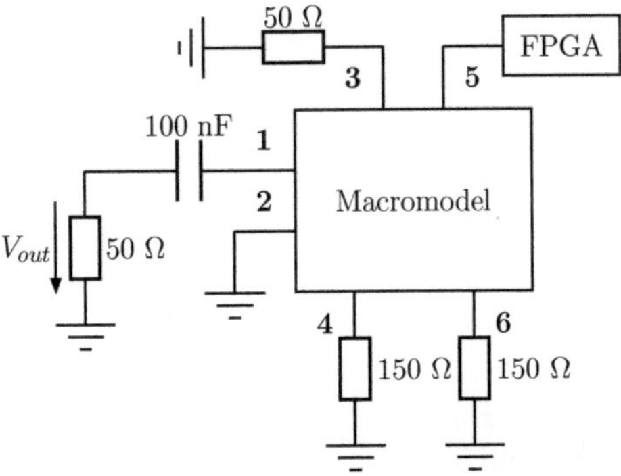

Figure 6.2: Circuit simulation setup of ESP example. The macromodel is the SPICE interpretation resulting from the reduced order model.

Again, we show the comparison between approximated and exact relative error in Figures 6.3 and 6.4 computed at 100 equally spaced frequency samples, respectively, for reduced systems of dimension 60 and 72. It can be seen that the residual-based estimation method introduced in Section 4.1.1 follows well the trend of the error. After 60 vectors, the convergence is not reached according to the approximated $\epsilon = 0.0023$. This value drops to $8.3 \cdot 10^{-6}$ beyond the threshold value $\epsilon_{th} = 10^{-4}$ after 72 vectors.

The computation of the transfer function took 22 min with the MOR code instead of 3h with MWS, which yields an acceleration factor of 8. This is due to the resonant behavior of the structure as can be seen in Figure 6.5 which shows the magnitude of $Z_{[1,1]}$. It should be stated that the time required for error estimation is 10 s. and thus less than 1% of the whole computation time. The performance comparison between MWS and MOR is also reported in Table 6.1.2.

Circuit Simulation

As the FPGA signal is available in frequency domain, we opted for an AC analysis in SPICE in the frequency range $[0 \text{ MHz}, \dots, 250 \text{ MHz}]$. As already explained in Section 3.6.2, the model obtained above is reduced a second time for SPICE model extraction. By doing so, we obtain a model of order 36 which is connected as illustrated in Figure 6.2. In Figure 6.6, we compare the spectra of the output voltage at the end of the wire obtained by measurement and simulation. It shows a good agreement in spite of the simplification we considered for the 3D model.

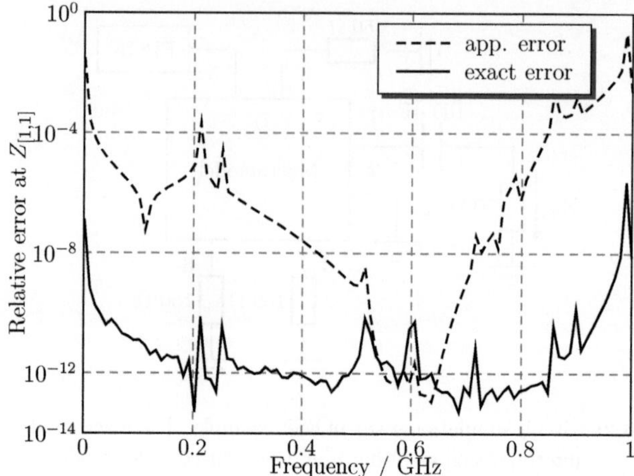

Figure 6.3: Approximated error vs. exact error at $Z_{[1,1]}$ of the ESP model for a reduced system of dimension 60.

	MWS®	MOR
method	FD (frequency sweep)	passive Padé
# unknowns	$650,000$	$650,000$
memory size	5.4 GB	4 GB
# inversions	24	1
accuracy	10^{-4}	10^{-4}
solver time	3 h	**22 min**

Table 6.1.1: Solver comparison between MOR and MWS for common analysis example

6.1.3 Analysis

The voltage at the end of the wire is relatively high and may lead to violation of EMC norms (CE). In order to improve this point by reducing the common mode signal magnitude, it is indispensable to analyze and find out the cause of the distortions. By identifying the sensitive parameters, EMC improving measures can be derived.

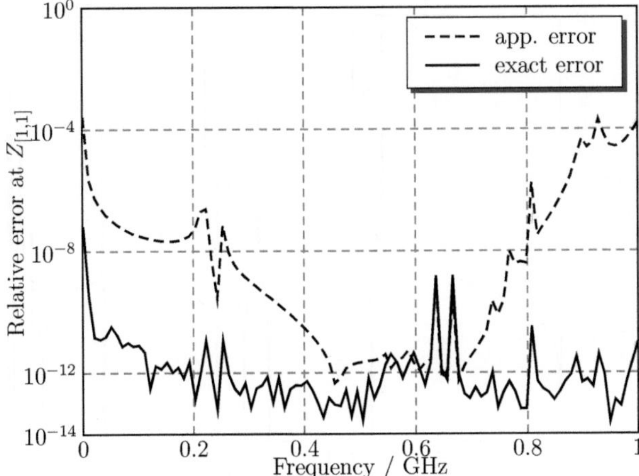

Figure 6.4: Approximated error vs. exact error at $Z_{[1,1]}$ of the ESP model for a reduced system of dimension 72.

We applied the method of EMC analysis introduced in Section 4.2 and considered only two ports:

- port 1 at the end of the wire

- port 5 at one end of the long trace,

whereas port 2 is shorted[3], and all other ports are open as they are not considered. The equivalent physical circuit should match with the 3D model up to 300 MHz, and thus consider the two first poles of the transfer function (Figure 6.5). By truncating the transfer function of the reduced model, we obtain a rational polynomial with 9 coefficients.

There were 4 proposed circuits which had the functional parts of the model in common, two inductors for the wires attached to the PCB ground which are coupled, and one for the signal trace. These differed in the position of parasitic capacitors. The circuit in Figure 6.7 is the only one which remained after the first selection[4]. In

---

[3]The second wire is connected to the measurement table.
[4]refer to Section 4.2.2

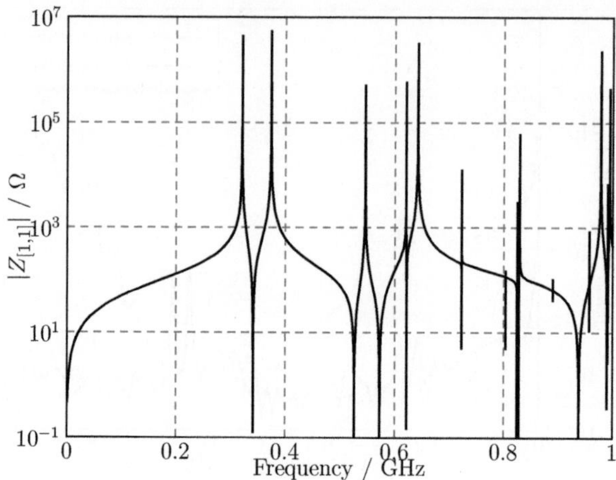

Figure 6.5: $Z_{[1,1]}$ of the ESP model computed with MOR.

fact, the transfer function of the other circuits did not have the same number of poles and zeros as the reference from MOR. Besides the functional elements, the circuit in Figure 6.7 consists of

- C_1 which models the capacitive coupling between PCB ground and table and also includes the coupling wire-table, enhanced by C_4,

- C_2 which corresponds to the coupling between the open end of the trace and PCB ground,

- and C_3 which models the capacitive coupling between trace and measurement table.

This leads to 8 unknowns which were computed through optimization following the method introduced in Section 4.2.

The result of a heuristic method[5] (based on a genetic algorithm performed with OptiSlang [96]) was set as guess of a gradient search for refinement. The objective function could be reduced to 0.007 and the transfer function of the circuit obtained

---

[5]They are suitable to find global optima, particularly for problems with constraints.

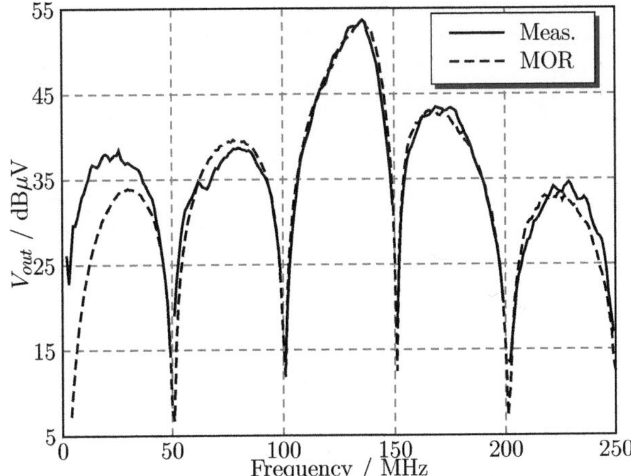

Figure 6.6: Comparison between measurements and MOR of the voltage at the end of the wire by an excitation at the long trace.

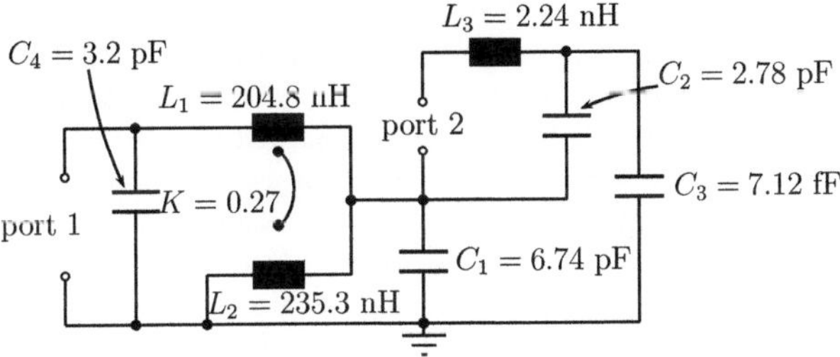

Figure 6.7: Equivalent physical circuit of the ESP model with main parasitic elements.

by optimization agree with MOR results as can be seen in Figures 6.8 and 6.9. Furthermore, the values retrieved are physically correct. In fact, L_1 and L_2 should be in the range of 200 nH as the wires are 20 cm long. It should be noted that the value L_2 is higher because the second wire is connected to the table via a PEC wall (Figure 6.1). Accordingly, the value of L_3 should be much smaller, and as C_1 also models the capacitive coupling between PCB ground and table, its value is greater than C_2. Furthermore, C_3 should be much smaller as the two other capacitors.

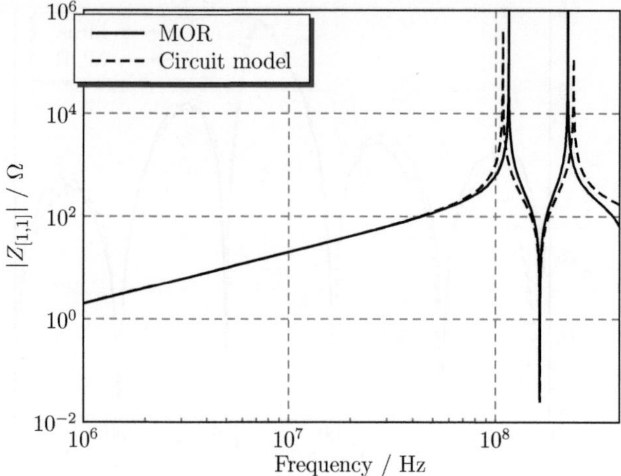

Figure 6.8: Magnitude of $Z_{[1,1]}$ resulting from MOR and equivalent circuit for the ESP model.

With a simple equivalent circuit which retrieves the couplings occurring on the structure, different sensitivity studies can be performed to identify the cause of the common mode distortions. In this way, it appeared that C_1 builds an LC-oscillator with L_1 and L_2 whose variations shift the frequency of the signal peak at 125 MHz in Figure 6.6 as its magnitude could be reduced by changing the value of the capacitive coupling between trace and table, C_3 which is in the range of fF. As illustrated in Figure 6.10, considering the short trace for excitation with an FPGA signal, the peak of the signal at the end of the wire could be reduced by more than 20 dBμV. Another improvement measure could be to shield the addressed trace.

6.2 DC/DC Converter

6.2.1 Model Description

The second device under test is a DC-DC converter. Its 3D model is illustrated in Figure 6.11. The whole simulation which is plugged in a SPICE environment should model a specific measurement setup for conducted emission which consists of

- a line impedance stabilization network (LISN) which is modeled in SPICE,

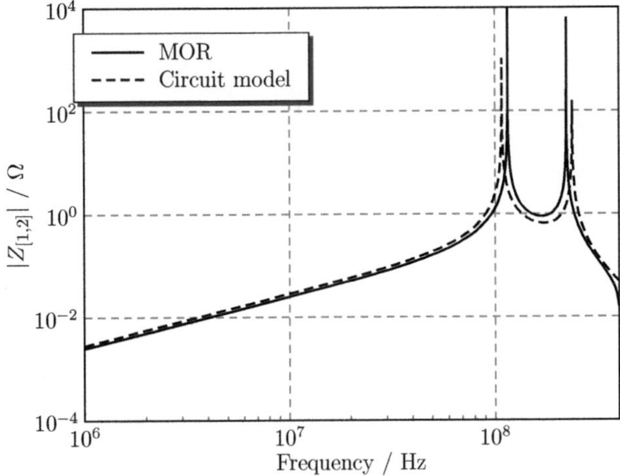

Figure 6.9: Magnitude of $Z_{[1,2]}$ resulting from MOR and an equivalent circuit for the ESP model.

- a wire model which connect the converter to the LISN, also modeled in SPICE

- the converter model with the filter rail which is modeled as 3D structure and computed with MOR,

- IGBT (insulated gate bipolar transistor) transistors which act as sources for the analysis,

- and the motor model obtained via measurement together with its connection cables modeled in SPICE.

The 3D model consists of the filter rail in its housing as the IGBTs are modeled with SPICE. The first 2 ports are connected to the LISN as the next 2 are linked to the IGBTs. The filter includes 3 capacitors:

- two C_y connected to the ground, and

- one C_x connecting the two filter rails.

The latter are also modeled as ports and replaced in the SPICE simulation by their equivalent circuits obtained through identification with measurement. The computational domain is bounded with PMC as radiation can be neglected[6]. The dicretization

---

[6]The metallic housing encloses the filter.

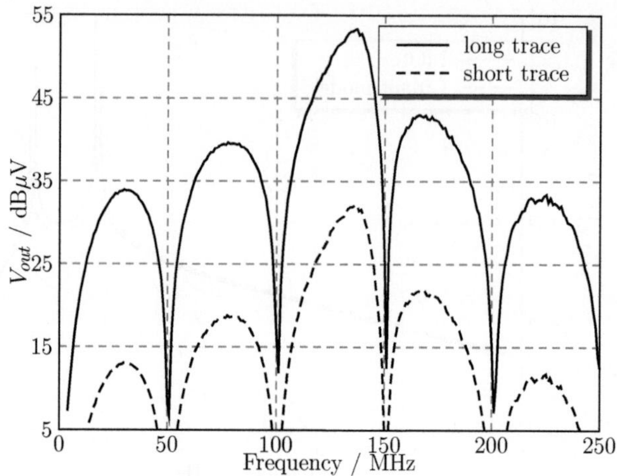

Figure 6.10: Comparison of distortion signals at end of wire for excitations at short and long trace.

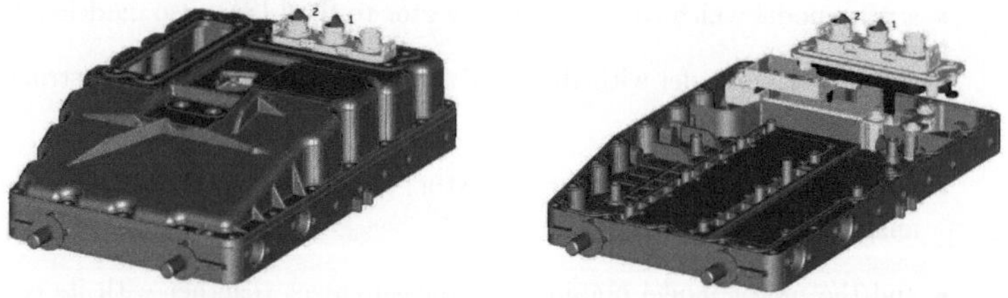

Figure 6.11: 3D model of the DC-DC converter. Left: with the housing cap on top. Right: the filter rail is put on evidence. The 4 ports are placed at the two connections to the LISN and to the IGBTs.

with MWS result in a system of $1.5 \cdot 10^6$ unknowns. The frequency range for the field simulation is $[1 \text{ MHz}, 100 \text{ MHz}]$.

6.2.2 Computation

The value to be computed is the voltage at the LISN. The computation of the whole measurement setup is performed in an SPICE environment following the concept illustrated in Figure 6.12.

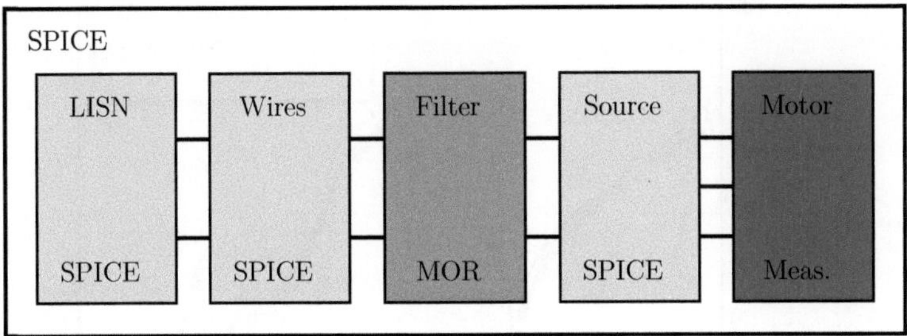

Figure 6.12: Simulation setup of the whole DC-DC converter system in SPICE environment.

As all other models are already available, two steps are performed:

- The first step aims to compute the transfer function of the 3D model and extract a macromodel.

- The second one consists of circuit simulation with consideration of all other models.

The transfer function computation of the 3D model with the MOR code requires 7.6 GB memory space and 16 min. instead of 43 min. with MWS. This corresponds to an acceleration factor of more than 2.5. This factor is so low because the transfer function is smooth in the considered frequency range. The frequency sweep of MWS thus needs only 7 frequency samples for convergence. The macromodel of order 21 after a second reduction step is plugged into the circuit simulation. Figure 6.13 shows the comparison of the voltage at the LISN between measurement and simulation. It can be seen that simulation agrees well with measurement with up to 10 dB discrepancy in the range [5 MHz, 20 MHz].

The main concern is the resonance at about 5.5 MHz where the magnitude is above EMC limits set by the motor client. It can be seen in Figure 6.13 that this resonance is matched by the simulation. The marginal computation time of the circuit simulation (3s.) made it possible to perform some coarse variations of the circuit elements in

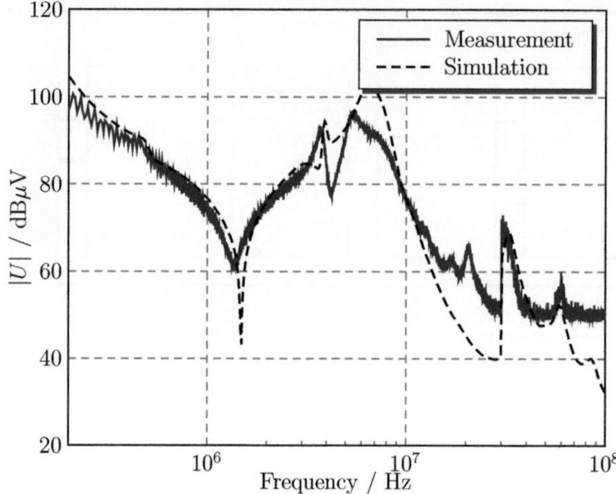

Figure 6.13: Comparison of the voltage at the LISN from measurement and simulation of the DC-DC converter.

order to understand the cause of this resonance. It thus appeared that the resonance is induced by common mode currents which flow through parasitic capacitors mainly on the motor side and are enhanced by the cable harness. This occurs also in combination with inductive and capacitive couplings occurring in the filter model. As the motor and its cable harness cannot be modified, the variation parameters in the optimization step were chosen in the 3D filter model.

6.2.3 Optimization

The geometry variations which were considered for optimization are

- location of capacitors along the filter rails,

- height of filter rails over housing,

- and distance of rails to each others.

The capacitors are considered as ports which are loaded with equivalent circuits afterwards in the SPICE environment. They have solely an influence on the right hand side matrix **B**. It should be noted that the capacitors in our optimization setup are directly linked to the housing and not through a shared metal conductor as in the original version. This metal conductor is thus modeled in the optimization setup by modifying the height of the rails over housing.

Variations of height and distance of filter rails to each other are applied by using the method introduced in Section 4.3.1. The 6 elements which are required to describe the rails are modified only in their y- and z-directions as they remain constant on the x-axis. The rails can get up to 5 mm close to each other and have a variation range in z of 17 mm. Each rail can have 8 different positions following the discretization which was set with a mesh length of 2.1 mm. The height varies between 7.5 mm and 15 mm above the housing with an increment of 1.3 mm. The capacitors are set according to the position of the rails.

The method we used for optimization is a multi-objective evolutionary algorithm presented in [108]. Of course any other optimization tool like OptiSlang [96] which we used for the analysis of the ESP example may be applied. As the goal was to reduce the magnitude of the voltage around the mentioned resonance, we considered the following objective function

$$f_{opt} = \sum_i |U(f_i)|, \text{ with } f_i \in [5 \text{ MHz}, 10 \text{ MHz}]. \tag{6.2.1}$$

We set 100 individuals for the start population, and the number of children after each generation to 2. The computation was stopped after 30 iterations as the results were not improving any more. The MOR computation was thus repeated 160 times and required 40 h. Apart from the acceleration factor of more than 2.5 compared to MWS, our method allows to save the meshing time of 2 min. after each modification of the 3D model in MWS environment.

The results of the optimization, illustrated in Figure 6.14, show an overall improvement of more than 32 dB. The optimal position of the rails is the lowest[7] one; by setting this height in the original setup we obtain more than 20 dB. This shows that the inductance of the capacitors connection to the housing is of great relevance. Further 10 dB are obtained by placing the capacitors optimally which has also an impact on capacitive and inductive couplings of the whole system.

---

[7]w.r.t. the height of rails above the housing.

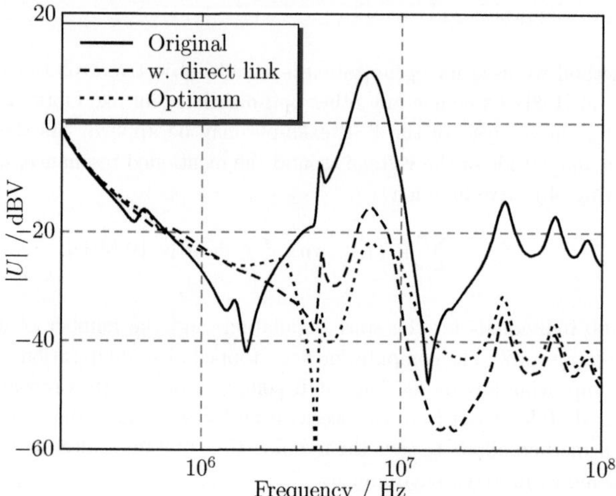

Figure 6.14: Improvement of the voltage at the LISN around 5 MHz after optimization.

7 Conclusion and Outlook

7.1 Conclusion

We proposed a method in this work to improve the efficiency of EMC simulation. For better integration in the development process, we considered the workflow consisting of EMC computation, analysis and optimization. MOR has been proved to be an essential part of this workflow.

The discretization with an orthogonal grid is performed for field computation with FIT, where it should be noted that the methods presented in this work can be extended to FEM matrices. The systems resulting from FIT preserve the passivity of the discretized structures except while considering PML boundaries which include layers with active behavior.

The consideration of MOR to compute the transfer function or generate macromodels of the discretized structures shows several advantages:

- reduced models,

- fast computation (acceleration factor of up to 30 compared to usual frequency sweep methods),

- appropriate for resonant structures,

- suitable for wide frequency ranges,

- high accuracy for macromodels,

- preservation of passivity.

The error control allows a reliable stop criterion and thus an optimal balance between accuracy and model size. Furthermore, the combination of modal truncation and Padé approximation improves the efficiency of MOR in presence of a high number of ports.

Considering more complex structures for simulation enables to reduce the modeling effort and to improve the accuracy and details of prediction. This occurs at cost of more computation complexity. In order to tackle this limit, a suitable parallelization technique for MOR has been implemented. It appeared that the multipoint-Padé approximation is not the best method as it is theoretically the case. In fact, spreading the generation of reduced models at different frequencies in a network and combining them afterwards is less efficient due to the poor orthogonalization performance.

The reduction method PRIMA with one interpolation point has thus been parallelized for computation on several processors. For this purpose, we used the library PETSc which is coupled with MPI and SuperLU as solver. The main computational bottle-neck, the LU decomposition of SuperLU, presents good scalability properties. It was possible to compute a complex problem with more than $8 \cdot 10^6$ unknowns on 48 processors in 1h30min. It should be noted that this could not be achieved on machines with 64 GB even by using iterative solvers after a week. The computation in time domain didn't converge also after two days even by using GPU because of the resonace behavior of the structure.

Another promising method to compute more complex structures has been presented. The Kronecker decomposition presented in this work is a compression method of matrices resulting from the discretization on orthogonal grids like FIT. In fact, in association with partial realization and an inverse approximation it potentially enables to handle systems in the order of 10^7 unknowns on common machines without requiring parallelization techniques.

Our proposed method for EMC analysis which consists on the generation of equivalent physical circuits could be successfully tested on the ESP model. It consists on matching the representations in rational polynomials of the transfer functions resulting from the MOR system and from the nodal analysis of some proposed circuits. It enables to automatically eliminate non-matching circuits among the proposed ones and compute the element values of the right circuit assuming it has been proposed. This method is essential for EMC analysis as it allows to identify the cause of parasitic effects occurring on the DUT.

EMC optimization consists of different measures which require on their turn the computation of different - mainly - geometric variants. A method to automatically run an optimization process with MOR in combination with a genetic algorithm was introduced. For this purpose, geometric variations are performed directly on the system matrices without requiring an interface to any mesh generator. By this way, the optimization is autonomous and fluent. On the other hand, we save the mesh generation time which may be in the range of some minutes per optimization step for complex structures.

The workflow consisting of EMC computation, analysis and optimization which considers MOR has been implemented. This enables to achieve the main goal of EMC engineering, which aims to provide an EMC optimized geometry in a more efficient way. This has been successfully applied to solve real-life problems. The conducted emissions of a DC-DC converter could be improved by 30 dB with our method.

7.2 Outlook

The parallelization method for MOR presented in this work enables to compute complex structures. Even though it allows faster computations, this solution has tremendous memory requirements which can only be met for models with more than 10^6 unknowns by using high performance clusters. Those facilities are very costly and not available to a lot of EMC engineers. The Kronecker decomposition presented in this work is a promising alternative. However, there is still research work to be done in order to provide a good approximation of the inverse which is indispensable for a satisfying convergence behavior of the partial realization method associated with it.

The EMC analysis method presented in this work requires a good guess for the equivalent circuits and thus relies on the experience of EMC engineers. In order to make it more reliable and available to a wider community, further works should enable to generate those circuits automatically. Starting with a given circuit involving the main functional elements, the method could heuristically add or remove parasitic elements and at the end retrieve the right circuit with the elements values.

The variations encountered in the EMC optimization part are of low rank. It is thus of great interest to implement a method which uses this property to fasten the optimization process. Particularly avoiding the inversion of the matrix after each modification should be the main task.

Finally, including MOR in well suited partitioning methods would be an optimal enabler for EMC simulation of CE and RE (radiated emissions). In fact, partitioning a whole domain would allow to handle the corresponding subdomains with the best suitable method. For CE measurement setups this could result in computing the PCB in its housing with MOR, stochastic cable harnesses with TLT (transmission line theory) methods and combining both together in a SPICE simulation to retrieve the voltages and currents at the LISN.

Bibliography

[1] DIN-VDE-Taschenbuch 515. *Elektromagnetische Verträglichkeit 1*. VDE-Verlag and Beuth Verlag, 1991.

[2] G. Durcansky. *EMV-gerechtes Gerätedesign*. Franzis, 3rd. edition, 1992.

[3] A. J. Schwab. *Elektromagnetische Verträglichkeit*. Springer Verlag, 1993.

[4] X. Tesche. *Electromagnetic Compatibility*. Springer Verlag, 1998.

[5] A. Englmaier. *Methoden und Modelle für die EMV-Simulation*. PhD thesis, Technische Universität München, 1998.

[6] S. M. Rao, D. R. Wilson, and A. W. Glisson. Electromagnetic Scattering by Surfaces of Arbritrary Shape. *IEEE Transactions on Antennas Propagation*, 30:409–418, 1982.

[7] J. Jin. *The Finite Element Method in Electromagnetics*. John Wiley & Sons, Inc., 2nd edition, 2002.

[8] K. S. Yee. Numerical Solution of Initial Boundary Value Problems Involving Maxwell's Equations in Isotropic Media. *IEEE Transactions on Antennas and Propagation*, 14(3):302–307, 1966.

[9] A. C. Antoulas. *Approximation of Large Scale Dynamical Systems*. Society for Industrial and Applied Mathematics, 2006.

[10] T. Wittig. *Zur Reduzierung der Modellordnung in elektromagnetischen Feldsimulation*. PhD thesis, TU Darmstadt, 2003.

[11] K. Krohne. *Order Reduction of Finite-Volume Models and its Application to Microwave Device Optimization*. PhD thesis, Swiss Federal Institute of Technology Zurich, 2007.

[12] O. Farle. *Ordnungsreduktionsverfahren für die Finite-Elemente-Simulation parameterabhängiger passiver Mikrowellenstrukturen*. PhD thesis, Universität des Saarlandes, 2007.

[13] G. Steinmair. *The Partial Element Equivalent Circuit Method and Model Order Reduction in Automative EMC Simulation.* PhD thesis, Johannes Kepler Universität, Linz, 2003.

[14] J. C. Maxwell. *A Treatise on Electricity and Magnetism,* volume 1 and 2. Oxford University Press, 1873.

[15] G. Lehner. *Elektromagnetische Feldtheorie für Ingenieure und Physiker.* Springer Verlag, Berlin/Heidelberg, 1996.

[16] T. Weiland. *Eine Methode zur Lösung der Maxwellschen Gleichungen für sechskomponentige Felder auf diskreter Basis.* Archiv für Elektronik und Übertragung, 1977.

[17] T. Weiland. *Elektromagnetisches CAD - Rechnergestützte Methoden zur Berechnung von Feldern.* Skriptum zur Vorlesung, TU Darmstadt, 2001.

[18] T. Weiland. Time Domain Electromagnetic Field Computation with Finite Difference Methods. *International Journal of Numerical Modeling,* pages 295–319, 1996.

[19] R. Schuhmann. *Die nichtorthogonale Finite-Integration-Methode zur Simulation elektromagnetischer Felder.* PhD thesis, TU Darmstadt, 1999.

[20] H. Krüger. *Zur numerischen Berechnung transienter elektromagnetischer Felder in gyrotropen Materialien.* PhD thesis, TU Darmstadt, 2000.

[21] B. Krietenstein, R. Schuhmann, P. Thoma, and T. Weiland. The Perfect Boundary Approximation Technique Facing the Big Challenge of High Precision Field Computation. *Proceedings of the XIX International Linear Accelerator Conference, Chicago, USA,* pages 860–862, 1998.

[22] D. Reinecke. *Zur Simulation dünner, perfekt elektrisch leitender Schichten durch die Methode der Finite Integration.* PhD thesis, Technische Universität Darmstadt, 2002.

[23] J.-P. Berenger. A Perfectly Matched Layer for the Absorption of Electromagnetic Waves. *J. Comput. Phys.,* 114(1):185–200, 1994.

[24] G. Mur. Absorbing Boundary Conditions for the Finite-Difference Approximation of the Time-Domain Electromagnetic-Field Equations. *IEEE Trans, Electromagn. Compat.,* 23(4):377–382, 1981.

[25] L. Zhao and A.C. Cangellaris. GT-PML: Generalized Theory of Perfectly Matched Layers and Its Application to the Reflectionless Truncation of Finite-Difference Time-Domain Grids. *IEEE MTT-S Int. Microwave Symp.*, 44(12):2555–2563, June 1996.

[26] W. C. Chew, J. M. Jin, and E. Michielssen. *Fast and Efficient Algorithms in Computational Electromagnetics*. Artech House, Boston, London, 2001.

[27] P. Thoma. *Zur numerischen Lösung der Maxwellschen Gleichungen im Zeitbereich*. PhD thesis, Technische Universität Darmstadt, 1997.

[28] M. Dohlus. *Ein Beitrag zur numerischen Berechnung elektromagnetischer Felder im Zeitbereich*. PhD thesis, TU Darmstadt, 1992.

[29] I. Munteanu and D. Iona. Some Properties of the Non-homogeneous Anisotropic Perfectly Matched Layer. *IEEE Trans. Magn.*, 34(5):2692–2695, September 1998.

[30] P. Hammes. *Zur numerischen Berechnung von Streumatrizen im Hochfrequenzbereich*. PhD thesis, TU Darmstadt, 1999.

[31] B. Trapp. *Zur numerischen Berechnung hochfrequenter elektromagnetischer Felder auf der Basis von Eigenlösungen*. PhD thesis, TU Darmstadt, 2002.

[32] P. Hahne. *Zur numerischen Berechnung zeitharmonischer elektromagnetischer Felder*. PhD thesis, Technische Universität Darmstadt, 1992.

[33] C. T. Chen. *Linear System Theory and Design*. Oxford University Press, New York, 1999.

[34] M. R. Wohlers. *Lumped and Distributed Passive Networks*. New York, Academics, 1969.

[35] B. D. O. Anderson and S. Vongpanidlerd. *Network Analysis and Synthesis*. Prentice Hall, Englewood Cliffs, New Jersey, 1973.

[36] R. W. Newcomb. *Linear Multiport Synthesis*. McGraw-Hill, New York, 1966.

[37] B. C. Moore. Principal Component Analysis in Linear System: Controllability, Observability and Model Reduction. *IEEE Transactions on Automatic Control*, 26:17–32, 1981.

[38] L. Pernebo and L. M. Silverman. Model Reduction via Balanced State Space Representation. *IEEE Transactions on Automatic Control*, 27(2):382–387, 1982.

[39] S. Gugercin and C. A. Antoulas. A Survey of Model Reduction by Balancing Truncation and some new Results. *International Journal of Control*, 77(8):748–766, 2004.

[40] M. Green. Balanced Stochastic Realizations. *Linear Algebra and Its Applications*, 98:211–247, 1988.

[41] A. C. Antoulas, D. C. Sorensen, and S. Gugercin. A Survey of Model Order Reduction Methods for Large-Scale Systems. Technical report, Rice University, December 2000.

[42] M. Dohlus, R. Schuhmann, and T. Weiland. Calculation of Frequency Domain Parameters using 3D Eigensolutions. *Int. J. Numer. Model.*, 12:41–68, 1999.

[43] Y. Saad. *Numerical Methods for Large Eigenvalue Problems*. Oxford Manchester, UK: Manchester University Press, 1992.

[44] L. T. Pillage and R. A. Rohrer. Asymptotic Waveform Evaluation for Timing Analysis. *IEEE Transactions on Computer-Aided Design of Integrated Circuits and Systems*, 9(4):352–266, 1990.

[45] R. Sanaie, E. Chiprout, M. S. Nakhla, and Q.-J. Zhang. A Fast Method for Frequency and Time Domain Simulation of High-speed VLSI Interconnects. *IEEE Transactions on Microwave Theory*, 42:2562–2571, 1994.

[46] E. Chiprout and M. S. Nakhla. Analysis of Interconnect Networks Using Complex Frequency Hopping (CFH). *IEEE Transactions on Computer-Aided Design of Circuits and Systems*, 14(2):186–200, 1995.

[47] Y. Saad. *Iterative Methods for Sparse Linear Systems*. SIAM Philadelphia, Pennsylvania, 2nd edition, January 2000.

[48] C. Lanczos. An Iteration Method for the Solution of the Eigenvalue Problem of Linear Differential and Integral Operators. *Journal of Research of the National Bureau of Standards*, 45(4):255–282, 1950.

[49] W. E. Arnoldi. *The Principle of Minimized Iteration in the Solution onf Eigenvalue Problem*. Quarterly of Applied Mathematics, 1951.

[50] P. Feldmann and R. W. Freund. Efficient Linear Circuit Analysis by Padé Approximation via the Lanczos Process. *IEEE Computer Society Press*, pages 170–175, 1994.

[51] B. N. Parlett. *The Symmetric Eigenvalue Problem.* Prentice Hall, Englewood Cliffs, 1980.

[52] E. J. Grimme. *Krylov Projection Methods for Model Reduction.* PhD thesis, Ohio State University, 1997.

[53] D. Skoogh. *An Implementation of a Parallel Rational Krylov Algorithm.* PhD thesis, Chalmer's University of Technology, Göteborg, 1996.

[54] K. Gallivan, E. J. Grimme, and P. van Dooren. A Rational Lanczos Algorithm for Model Reduction. *Numerical Algorithm*, 12:33–63, 1996.

[55] D. Skoogh. A Rational Krylov Method for Model Order Reduction. Technical Report 1998-47, Department of Mathematics, Chalmers University of Technology and the University of Göteborg, 1998.

[56] R. D. Slone, R. Lee, and J. F. Lee. Well-Conditioned Asymptotic Waveform Evaluation for Finite Elements. *IEEE Transactions on Antennas and Propagation*, 51(9):2442–2447, 2003.

[57] R. W. Freund. Krylov-Subspace Methods for Reduced-Order Modeling in Circuit Simulation. *Journal of Computational and Applied Mathematics*, 123:395–421, 2000.

[58] T. Zhou, S. L. Dvorak, and J.L. Prince. Application of the Padé via Lanczos (PVL) Algorithm to Electromagnetic Systems with Expansion at Infinity. In *Proc. of 50th Electronic Components and Technology Conference*, pages 1515–1520, 2000.

[59] A. Odabasioglu, M. Celik, and L. T. Pileggi. PRIMA: Passive Reduced-Order Interconnect Macromodeling Algorithm. *IEEE Transactions on CAD of Integrated Circuits and Systems*, 17(8):645–654, August 1998.

[60] R. W. Freund and P. Feldmann. Reduced-Order Modeling of Large Linear Passive Multi-Terminal Circuits Using Matrix-Padé Approximation. *Proceedings of the Design Automation and Test Conference in Europe, IEEE Computer Society Press*, pages 530–537, 1998.

[61] J. Phillips, L. Daniel, and L. Silveira. Guaranteed Passive Balancing Transformations for Model Order Reduction. *IEEE Transactions on CAD of Integrated Circuits and Systems*, 22(8):1027–1041, Aug. 2003.

[62] N. Wong, V. Balakrishnan, and C.-K. Koh. Passivity-Preserving Model Reduction via a Computationally Efficient Project and Balanced Scheme. *Conference on Design Automation*, 41:369–374, 2004.

[63] CST GmbH. *MICROWAVE STUDIO*. Bad Nauheimer Strasse 19, 64289 Darmstadt.

[64] L. W. Nagel. A computer program to simulate semiconductor circuits. Technical Report ERL-M520, Univ. California, Berkeley, May 1975.

[65] M. Elzinga, K. Virga, and J. Prince. Improved Gobal Rational Approximation Macromodeling Algorithm for Transient Simulation of Interconnects. Technical report, Department of Electrical and Computer Engineering, University of Arizona, 1999.

[66] B. Gustavsen and A. Semlyen. A Robust Approach for System Identification in the Frequency Domain. *IEEE Transactions on Power Delivery*, 19(3):1167–1173, July 2004.

[67] S. Grivet-Talocia and A. Ubolli. On the Generation of Large Passive Macromodels for Complex Interconnect Structures. *IEEE Transactions on Advanced Packaging*, 29(1):39–54, February 2006.

[68] S. Grivet-Talocia and M. Bandinu. Improving the Convergence of Vector Fitting for Equivalent Circuit Extraction from Noisy Frequency Responses. *IEEE Transactions on Electromagnetic Compatibility*, 48(1):104–120, February 2006.

[69] B. Gustavsen and A. Semlyen. Enforcing Passivity for Admittance Matrices Approximated by Rational Functions. *IEEE Transactions on Power Systems*, 16(1):97–104, February 2001.

[70] S. Grivet-Talocia. Passvity Enforcement via Pertubation of Hamiltonian Matrices. *IEEE Transactions on Circuits Systems*, 51(9):1755–1769, September 2004.

[71] C. W. Ho, A. E. Ruehli, and P. A. Brennan. The Modified Nodal Approach to Network Analysis. *IEEE Transactions on Circuits and Systems*, 22:504–509, 1975.

[72] V. Raghavan, J. E. Bracken, and R. A. Rohrer. AWESpice: A General Tool for the Accurate and Efficient Simulation of Interconnect Problems. *IEEE/ACM International Conference on DAC*, pages 87–92, June 1992.

[73] S. Y. Kim, N. Gopal, and L. T. Pillage. Time-domain macromodels for vlsi interconnect analysis. *IEEE Transactions on Computer-Aided Design*, 13:1257–170, October 1994.

[74] T. L. Quarles. The SPICE3 Implementation Guide. Technical Report ERL-M89/44, Univ. California, Berkeley, 1989.

[75] A. Odabasioglu, M. Celik, and L. T. Pileggi. Practical Considerations for Passive Reduction of RLC Circuits. *IEEE/ACM International Conference on CAD*, pages 214–219, 1999.

[76] P. Heres. *Robust and Efficient Krylov Subspace Methods for Model Order Reduction*. PhD thesis, TU Eindhoven, December 2005.

[77] T. Tischler. *Die Perfectly-Matched-Layer-Randbedingung in der Finite-Differenzen-Methode im Frequenzbereich: Implementierung und Einsatzbereiche*. PhD thesis, TU Berlin, 2004.

[78] P. Feldmann. Model Order Reduction Techniques for Linear Systems with Large Numbers of Terminals. *Proceedings of the Design, Automation and Test in Europe Conference and Exhibition*, 2:944– 947, Februar 2004.

[79] P. Li and W. Shi. Model Order Reduction of Linear Networks with Massive Ports via Frequency-Dependent Port Packing. In *Proc. of 43rd annual Design Automation Conference, San Antonio, CA, USA*, pages 267–272, July 2006.

[80] B. Yan, L. Zhou, S. X.-D. Tan, J. Chen, and Bruce McGaughy. DeMOR: Decentralized Model Order Reduction of Linear Networks with Massive Ports. In *Proc. of 45th annual Design Automation Conference, Anaheim, CA, USA*, pages 409–414, June 2008.

[81] J. Stoev and J. Moehring. Model Reduction Approaches for MIMO Systems: Combining Krylov Methods and Modal Reduction. *VDI/VDE-GMA-Fachausschuss "Modellbildung, Identifikation und Simulation in der Automatisierungstechnik", Salzburg, Austria*, September 2008.

[82] M. J. Flynn. Some Computer Organization and their Effectiveness. *IEEE Transactions on Computers*, C-21(9):948–960, 1972.

[83] A. S. Tanenbaum. *Computerarchitektur*. Pearson Studium, 5th edition, 2006.

[84] OpenMP Architecture Review Board. *OpenMP Application Program Interface*. http://www.openmp.org, 2005.

[85] Top Computing Sites. http://www.top500.0rg/.

[86] Message Passing Interface Forum. *MPI: A Message-Passing Interface Standard.* http://www-unix.mcs.anl.gov/mpi, 1994.

[87] M. J. Quinn. *Parallel Programming in C with MPI and OpenMP.* Mc Graw Hill, 2003.

[88] G. M. Amdahl. Validity of the Single Processor Approach to Achieving Large Scale Computing Capabilities. *AFIPS Conference Proceedings National Computing Conference*, pages 483–485, 1967.

[89] S. Balay, K. Buschelman, W. D. Gropp, D. Kaushik, M. G. Knepley, L. C. McInnes, B. F. Smith, and H. Zhang. PETSc Web page, 2009. http://www.mcs.anl.gov/petsc.

[90] S. Balay, K. Buschelman, V. Eijkout, W. D. Gropp, D. Kaushik, M. G. Knepley, L. C. McInnes, B. F. Smith, and H. Zhang. PETSc Users Manual. Technical Report ANL-95/11-Revision 3.0.0, Argonne National Laboratory, 2008.

[91] E. Anderson, Z. Bai, C. Bischof, S. Blackford, J. Demel, J. Dongarra, J. Du Croz, A. Greenbaum, S. Hammarling, A. McKenney, and D. Sorensen. *LAPACK Users' Guide.* Society for Industrial and Applied Mathematics, Philadelphia, PA, 3rd edition, 1999.

[92] M. Henzler. *Parallelisierung eines Codes zur Modellordnungsreduktion.* Bachelor's thesis, University of Applied Sciences Esslingen, 2010.

[93] D.J. Tylavsky and G.R.L. Sohie. Generalization of the Matrix Inversion Lemma. *Proceedings of IEEE*, 74(7):1050–1052, 1986.

[94] O. Schenk. *Scalable Parallel Sparse LU Factorization Methods on Shared Memory Multiprocessors.* PhD thesis, ETH Zürich, 2000.

[95] L. Grasedyck, W. Hackbusch, and R. Kriemann. Performance of H-LU Preconditioning for Sparse Matrices. *Computational Methods in Applied Mathematics*, 8(4):336–349, 2008.

[96] OptiSlang. *The Optimizing Structural Language, Version 3.0.1.* DYNARDO, Weimar, 2009.

[97] L. Greengard and V. Rokhlin. A Fast Algorithm for Particle Simulation. *Journal of Computational Physics*, 73:325–348, 1987.

[98] W. Hackbusch and Z. P. Nowak. On the Fast Matrix Multiplication in the Boundary Element Method by Panel Clustering. *Computing*, 54:463–491, 1989.

[99] W. Hackbusch. A Sparse Matrix Arithmetic based on Sparse Matrices. Part I: Introduction to H-Matrices. *Computing*, 62:89–108, 1999.

[100] S. Kurz, O. Rain, and S. Rjasanow. The Adaptive Cross-Approximation Technique for the 3D-Boundary-Element Method. *IEEE Transactions on Magnetics*, 38:421–424, 2002.

[101] M. Bebendorf and S. Rjasanow. Adaptive Low Rank Approximation for Collocation Matrices. *Computing*, 70:1–23, 2003.

[102] I. Ibraghimow. Application of the Three-Way Decomposition for Matrix Compression. *Numer. Linear Algebra Appl.*, pages 1–16, 2002.

[103] L. de Lathauwer, B. de Moor, and J. Vandewalle. A Multilinear Singular Value Decomposition. *SIAM J. Matrix Anal. Appl.*, 21(4):1253–1278, 2000.

[104] L. de Lathauwer, B. de Moor, and J. Vandewalle. On The Best Rank-1 and Rank-(R1,R2, . . . , RN) Approximation of Higher-Order Tensors. *SIAM J. Matrix Anal. Appl.*, 21(4):1324–1342, 2000.

[105] I. Ibraghimow. The Parallel Decomposition. *Numer. Linear Algebra Appl.*, 2007.

[106] I. Ibraghimow. Sublinear Complexity of Iterative Methods for the Kroneker Product Matrices. *Numer. Linear Algebra Appl.*, 2007.

[107] I. V. Oseledets and E. E. Tyrtyshnikov. Approximate Inversion of Matrices in the Process of Solving a Hypersingular Integral Equation. *Computational Mathematics and Mathematical Physics*, 2(2):302–313, 2005.

[108] M. N. Albunni. Multiobjective Optimization of the Design of Electrical Machines Using Evolutionary Algorithms. Master's thesis, Universität Bremen, 2005.

List of Figures

List of Tables

Abbreviations

3D	three dimension, three dimensional
AWE	asymptotic waveform evaluation
BEM	boundary element method
BIBO	bounded-input, bounded-output
CISPR	comité international spécial pour les pertubations radioélectriques
CENELEC	comité européen de normalisation electrotechnique
CE	conducted emissions
CFH	complex frequency hopping
DC	direct current
DEK	Deutsches Elektrotechnisches Komitee
DMM	distributed memory machine
DOF	degree of freedom
DUT	device under test
EMC	electromagnetic compatibility
ESP	electronic stabilization program
EVD	eigenvalue decomposition
FD	finite difference
FEM	finite element method
FIT	finite integration technique
FPGA	field programable gate array
FVM	finite volume method
GT-PML	generalized perfectly matched layer theory
HF	high frequency
IC	integrated circuit
IGBT	insulated gate bipolar transistor
ILU	incomplete LU decomposition
KCL	Kirchhoff's current law
LISN	line impedance stabilization network
LTI	linear time invariant
MIMO	multiple-input multiple-output
MNA	modified nodal analysis

MOR	model order reduction
MPI	message passing interface
MWS	microwave studio
PBA	perfect boundary approximation
PCB	printed circuit board
PEC	perfect electric condition
PMC	perfect magnetic condition
PML	perfectly matched layers
POD	proper orthogonal decomposition
PRIMA	passive reduced interconnect macromodeling
PVL	Padé via Lanczos
RAM	random access memory
RE	radiated emissions
RGA	relative gain array
ROC	region of convergence
SIMO	single-input multiple-output
SISO	single-input single-output
SMM	single memory machine
SVD	singular value decomposition
TE	transversal electric
TM	transversal magnetic
TSL	two-step Lanczos
TST	thin sheet technique
WCAWE	well-conditioned AWE

Acknowledgments

This thesis is the accomplishment of many years of efforts and would have not been possible without the assistance of many people. I would like to thank them by this way.

First of all, I would like to thank Prof. Dr.-Ing. Thomas Weiland for giving me the opportunity to work in his institute, for his disponibility and advices during the last years. I thank also Prof. Dr. techn. Romanus Dyczij-Edlinger for accepting to be my second reviewer.

I would like to express my gratitude to the company Robert Bosch GmbH who sponsored my thesis and gave me the opportunity to work in a competitive environment and on challenging topics. Especially, Dr.-Ing. Jan Hansen, my adviser, for his commitment and the helpful advices, my group managers Dr. Christian Waldschmidt and Dr. Thomas Fritzsche, Dr. Christoph Keller, Hermann Aichele and Markus Gonser for the fruitful discussions and my colleagues from CR/ARE1 for the positive atmosphere.

My adviser at TEMF, Dr.-Ing. Wolfgang Ackermann, deserves my gratitude for his willing to share his huge knowledge with me. Thanks also to my colleagues from TEMF especially for the "instructing" KWT evenings.

Furthermore I would like to thank my parents, brothers, sisters and friends for their moral support. Especially my father who didn't spare any financial and moral effort to assist me and who continues to inspire me until now. "Je vous remercie tous du fond du coeur".

Last but not least, I would like to express my extreme gratitude to my wife Sandrine and our sunshine Kelyan who directly felt my stress especially in the last stage of my thesis. Your patience and your love have been precious to me.

Curriculum Vitae

24. Mai 1982	geboren in Douala, Kamerun
1987 - 1991	Grundschule in Douala
1991 - 1998	Gymnasium in Douala
Juni 1998	Abitur mit Schwerpunkt Mathematik und Physik
2001 - 2006	Studium der Elektrotechnik an der TU Berlin
2003 - 2006	Tutor für Grundlagen der Elektrotechnik an der TU Berlin
Juni 2006	Diplomarbeit "Bit Error Rate Estimation for 16-QAM Systems in Radio over Fiber Links"
seit März 2007	Externer Doktorand am Institut Theorie elektromagnetischer Felder des Fachbereichs für Elektrotechnik und Informationstechnik der TU Darmstadt
seit Feb. 2010	Entwicklungsingenieur bei der Robert Bosch GmbH